TIMBER DESIGN

for the Civil and Structural Professional Engineering Exams

Sixth Edition

Robert H. Kim, MSCE, PE
Jai B. Kim, PhD, PE

Professional Publications, Inc.
Belmont, CA

How to Locate Errata and Other Updates for This Book

At Professional Publications, we do our best to bring you error-free books. But when errors do occur, we want to make sure that you know about them so they cause as little confusion as possible.

A current list of known errata and other updates for this book is available on the PPI website at **www.ppi2pass.com**. From the website home page, click on "Errata." We update the errata page as often as necessary, so check in regularly. You will also find instructions for submitting suspected errata. We are grateful to every reader who takes the time to help us improve the quality of our books by pointing out an error.

National Design Specification® and NDS® are registered trademarks of the American Forest & Paper Association, American Wood Council, 1111 19th Street N.W., Suite 800, Washington, D.C. 20036.

Timber Design for the Civil and Structural Professional Engineering Exams
Sixth Edition

Current printing of this edition: 1

Printing History

edition number	printing number	update
5	1	New edition. Copyright update.
5	2	Minor corrections.
6	1	New edition. Copyright update.

Printed in the United States of America

Professional Publications, Inc.
1250 Fifth Avenue, Belmont, CA 94002
(650) 593-9119
www.ppi2pass.com

Library of Congress Cataloging-in-Publication Data
Kim, Robert H., 1964-
   Timber design for the civil and structural professional engineering exams / Robert H.
Kim, Jai B. Kim.--6th ed.
      p. cm.
   Includes index.
   ISBN 1-888577-86-X
   1. Building, Wooden--United States--Examinations--Study guides. 2. Building,
Wooden--United States--Problems, exercises, etc. I. Kim, Jai B., 1934- II. Title.

TA666.K56 2003
624.1′84--dc21
                                                    2003046577

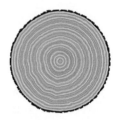

Table of Contents

Chapter 13
Practice Problems

Preface

This book is intended to be used by engineers preparing for the civil and structural professional engineering examinations, and as a reference book for practicing engineers.

One of the main purposes of this book is to acquaint the reader with timber engineering principles as thoroughly as possible and in the shortest time possible. Thus, in-depth theory about structural timber is not covered in this book, and the reader is urged to consult the publications listed in the References section for more comprehensive explanations of timber theory. It is necessary for users of this book to have a basic understanding of mechanics of materials.

Both the civil and structural professional engineering examinations may include timber engineering problems. Most of this book extensively references the 1997 edition *ANSI/AF&PA NDS-1997 National Design Specification for Wood Construction* (NDS), copyright 1997; the 1997 edition of *Supplement, NDS for Wood Construction*, copyright 1997; and the September 1998 Errata and April 2002 Errata/Addendum to the 1997 edition of the NDS. It is essential that users of this book have these NDS publications to clearly understand structural timber design principles.

The Plywood section of this book, Ch. 12, contains all necessary data for working with plywood as given in *Plywood Design Specification* (PDS), January 1997, copyright 1998; *PDS Supplement #2: Design and Fabrication of Plywood-Lumber Beams* (PDS Supplement Two), July 1992, copyright 1992; and *Design/Construction Guide: Diaphragms and Shear Walls*, copyright 1997.

The reader should be aware that while the NDS is incorporated into many major building codes and structural timber specifications, there may be codes and specifications that differ from, and take priority over, the specifications in the NDS.

While we carefully checked and rechecked the contents of this book, it is inevitable that some mistakes were passed over. We appreciate readers' feedback regarding any errors or ambiguities that may be found in this book, along with questions and comments. You may view or submit errata through PPI's website at **www.ppi2pass.com**.

We hope that this book will provide valuable information for the professional engineering examinations and that it will continue to serve the reader as a useful reference in the years to follow. Good luck.

Robert H. Kim, MSCE, PE
Jai B. Kim, PhD, PE

Acknowledgments

We are grateful to the publisher, Professional Publications, Inc., and are especially indebted to Aline Magee (Acquisitions Editor), Cathy Schrott (Production Manager), Jessica R. Holden (Project Editor and Proofreader), and Kate Hayes (Typesetter) for their continued support. We also thank Professor Keith F. Faherty of Marquette University for ready responses and discussions on timber engineering principles; Mr. Drew R. Potts, PE, of Watsontown, Pennsylvania, for providing a much needed review of the manuscript for this book; and Jacklyn Bowen of Elk Grove, California, and Robert H. Mayer, Associate Professor of Ocean Engineering at the U.S. Naval Academy, for their comments on the previous edition.

We are grateful too for the painstaking and careful review of this sixth edition by Professor Thomas H. Miller, PhD, PE, Associate Professor of Civil Engineering at Oregon State University.

We also wish to acknowledge Bucknell University in Lewisburg, Pennsylvania, for providing a conducive environment for the writing of this book, and the students in our CENG 403 Wood Engineering course at Bucknell for their comments, corrections, and suggestions.

Finally, we express gratitude to Mrs. Yung J. Kim, mother of Robert and wife of Jai, who provided the necessary support and motivation for the writing of this book.

How to Use This Book

It is essential that this book be used with the 1997 NDS and 1997 NDS Supplement cited in the References section. Since the civil and structural professional engineering examinations are open book, it is highly recommended that, prior to the examination, you mark pages in both the NDS and this book that have critical information, such as tables and the more commonly used equations, for ready reference during the exam.

This book references the applicable NDS publication prior to discussions of design principles and presentation of the example problems for each topic. A photocopy of the NDS and NDS Supplement can be purchased from the American Wood Council Publications Department, (800) 890-7732.

Although Ch. 12 includes the necessary information and data for working with plywood, you may purchase *Plywood Design Specification, PDS Supplement #2: Design and Fabrication of Plywood-Lumber Beams* and *Design/Construction Guide: Diaphragms and Shear Walls* from the American Plywood Association (www.apawood.org).

After covering each topic and looking up the NDS references where given, you are encouraged to carefully follow the example problems and then, without referring back, duplicate the solutions.

Keep in mind that in actual design situations there are often several correct solutions to the same problem, so it may not be constructive to persist at a problem until the exact solution is duplicated. If you decide to use this book as a primary reference during the exam, we recommend that you become as familiar with it, and any other primary reference, as possible.

It is essential that you bring a calculator to the exam. You are not permitted to share books, reference materials, calculators, or any other items with other candidates during the exam. Since you are usually not permitted to plug a calculator into an electrical outlet during the exam, be sure that your calculator batteries are fully charged prior to entering the exam room. Be careful not to violate exam rules, since disqualification from an exam may jeopardize your eligibility for future registration.

During the exam, we suggest that you work the easier problems first to maximize the number of points you can earn. It is not recommended that exam candidates leave exams early. If any time remains after you've completed the exam, recheck your calculations and answers.

After studying this book, you should be able to solve most common structural timber problems, both on the exams and in real design applications. Good luck on the exams.

List of Tables

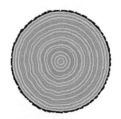

List of Figures

Nomenclature

Unless defined otherwise in the text, the following symbols are used in this book.

symbol	definition (units)
A	area (in^2, ft^2)
A_m	gross cross-sectional area of main wood member (in^2)
A_n	net cross-sectional area of a tension or compression member at a connection (in^2)
A_s	sum of gross cross-sectional areas of side member(s) (in^2)
b	width or thickness of rectangular beam cross section (in)
c	buckling and crushing interaction factor for columns
c	distance between neutral axis and extreme fiber (in, ft)
C_b	bearing area factor
C_d	penetration depth factor for connections
C_D	load duration factor
C_{di}	diaphragm factor for nail connections
C_{eg}	end grain factor for connections
C_f	form factor for bending stress
C_F	size factor for sawn lumber
C_{fu}	flat use factor for bending stress
C_g	group action factor for connections
C_H	shear stress adjustment factor
C_L	beam stability factor
C_M	wet service factor
C_P	column stability factor
C_r	repetitive member factor (bending stress) for dimension lumber

symbol	definition (units)
C_{st}	metal side plate factor for 4 in shear plate connections
C_t	temperature factor
C_T	buckling stiffness factor for 2×4 and smaller dimension lumber in trusses
C_{tn}	toe-nail factor for nail connections
C_V	volume factor for glulam
C_Δ	geometry factor for connections
d	width or depth of rectangular beam cross section (in)
d	pennyweight of nail or spike
D	fastener diameter (in)
d_x	width of rectangular column parallel to y-axis, used to calculate column slenderness ratio about x-axis (in)
d_y	width of rectangular column parallel to x-axis, used to calculate column slenderness ratio about y-axis (in)
E, E'	tabulated and allowable modulus of elasticity (lbf/in^2)
E_{axial}	modulus of elasticity of glulam for axial deformation calculation (lbf/in^2)
E_m	modulus of elasticity of main member (lbf/in^2)
E_s	modulus of elasticity of side member (lbf/in^2)
E_x	modulus of elasticity about x-axis (lbf/in^2)
E_y	modulus of elasticity about y-axis (lbf/in^2)
f_b	actual bending stress (lbf/in^2)

symbol	definition (units)
F_b, F_b'	tabulated and allowable bending stress (lbf/in^2)
F_b^*	tabulated bending stress multiplied by all applicable adjustment factors except C_L (lbf/in^2)
F_b^{**}	tabulated bending stress multiplied by all applicable adjustment factors except C_V (lbf/in^2)
F_{bE}	critical buckling (Euler) value for bending member (lbf/in^2)
f_{bx}	actual bending stress about strong (x) axis (lbf/in^2)
F_{bx}, F_{bx}'	tabulated and allowable bending stress about strong (x) axis (lbf/in^2)
$F_{bx,c/t}$	tabulated bending stress about the x-axis with compression zone stressed in tension
$F_{bx,c/t}'$	allowable bending stress about the x-axis with compression laminations stressed in tension
$F_{bx,t/t}$	tabulated bending stress about the x-axis tension zone stressed in tension
$F_{bx,t/t}'$	allowable bending stress about the x-axis with high-quality tension laminations stressed in tension
f_{by}	actual bending stress about weak (y) axis (lbf/in^2)
F_{by}, F_{by}'	tabulated and allowable bending stress about weak (y) axis (lbf/in^2)
f_c	actual compression stress parallel to grain (lbf/in^2)
$f_{c\perp}$	actual compression stress perpendicular to grain (lbf/in^2)
F_c, F_c'	tabulated and allowable compression stress parallel to grain (lbf/in^2)
F_c^*	tabulated compression stress parallel to grain multiplied by all applicable adjustment factors except C_P (lbf/in^2)
F_{cE}	critical buckling (Euler) value for compression member (lbf/in^2)
$F_{c\perp}, F_{c\perp}'$	tabulated and allowable compression stress perpendicular to grain (lbf/in^2)
F_e	dowel bearing strength (lbf/in^2)
$F_{e\parallel}$	dowel bearing strength parallel to grain for bolt or lag bolt connection (lbf/in^2)

symbol	definition (units)
$F_{e\perp}$	dowel bearing strength perpendicular to grain for bolt or lag bolt connection (lbf/in^2)
F_{em}	dowel bearing strength of main member (lbf/in^2)
F_{es}	dowel bearing strength of side member (lbf/in^2)
$F_{e\theta}$	dowel bearing strength at angle to grain θ for bolt or lag bolt connection (lbf/in^2)
f_g	actual bearing stress parallel to grain (lbf/in^2)
F_g, F_g'	tabulated and allowable bearing stress parallel to grain (lbf/in^2)
F_s	tabulated plywood rolling shear stress (lbf/in^2)
f_t	actual tension stress in a member parallel to grain (lbf/in^2)
F_t, F_t'	tabulated and allowable tension stress parallel to grain (lbf/in^2)
F_u	ultimate tensile strength for steel (lbf/in^2, k/in^2)
f_v	actual shear stress parallel to grain (horizontal shear) in a beam (lbf/in^2)
F_v, F_v'	tabulated and allowable shear stress parallel to grain (horizontal shear) in a beam (lbf/in^2)
F_{yb}	bending yield strength of fastener (lbf/in^2, k/in^2)
f_θ	actual bearing stress at angle, θ, to grain
F_θ'	allowable bearing stress at angle to grain θ (lbf/in^2)
G	specific gravity
h	width or depth of rectangular beam cross section (in)
I	moment of inertia (in^4)
K	correction factor for effective section moduli for plywood
K_{bE}	Euler buckling coefficient for beams
K_{cE}	Euler buckling coefficient for columns
K_D	diameter coefficient for nail and spike connections

symbol	definition (units)		symbol	definition (units)
K_e	effective length factor for column end conditions (buckling length coefficient for columns)		P	total applied concentrated load or total axial load (lbf)
K_L	loading condition coefficient for evaluating volume effect factor C_V for glulam beams		P, P'	nominal and allowable lateral design value parallel to grain for a single split ring connector unit or shear plate connector unit (lbf)
KS	effective section moduli for plywood rather than I/C (in^3)		P_{allow}	total allowable concentrated load or total allowable axial load (lbf)
K_θ	angle to grain coefficient for bolt and lag bolt connections		Q	statical moment of an area about the neutral axis (in^3)
l	length (in)		Q, Q'	nominal and allowable lateral design value perpendicular to grain for a single split ring connector unit or shear plate connector unit (lbf)
l	length of column (in)			
L	length of beam between points of zero moment (ft)			
L	span length of beam (in, ft)		Q_{allow}	total allowable concentrated load or total allowable axial load (lbf)
L	spike length (in)		R	reaction force (lbf)
l_b	bearing length (in)		R_B	slenderness ratio of laterally unbraced beam
L_c	cantilever length in cantilever beam system (ft)			
l_e	effective unbraced length of column (in)		s	fastener spacing (in)
l_e	effective unbraced length of compression side of beam (in)		S	section modulus (in^3)
			t	thickness (in)
l_e/d	slenderness ratio of column		T	tension force (lbf, k)
$(l_e/d)_x$	slenderness ratio of column for buckling about strong (x) axis		t_m	thickness of main (thicker) member (in)
$(l_e/d)_y$	slenderness ratio of column for buckling about weak (y) axis		t_s	effective thickness for shear (in)
			t_s	thickness of side (thinner) member (in)
l_m	length of bolt in wood main member (in)		t_{washer}	thickness of washer (in)
l_s	total length of bolt in wood side member(s) (in)		$t_\parallel$	effective thickness of parallel plies for plywood (in)
l_u	laterally unbraced length of compression side of beam or unbraced length of column (in)		v	unit shear stress (lbf/ft, lbf/in, lbf/in^2)
			V	shear force (lbf)
l_x	unbraced length of column considering buckling about strong (x) axis (in)		V_h	horizontal shear force (plywood) (lbf)
l_y	unbraced length of column considering buckling about weak (y) axis (in)		V_s	rolling shear force (plywood) (lbf)
M	bending moment (in-lbf, in-k, ft-lbf, ft-k)		w, W	uniformly distributed and total force (lbf/in, lbf/ft, lbf/ft^2, k/ft, k/in, k/ft^2, lbf, k)
N, N'	nominal and allowable lateral design value at angle to grain θ for a single split ring or shear plate connector (lbf)		W, W'	nominal and allowable withdrawal design value for a single fastener (lbf, lbf/in)
			W_{allow}	allowable uniform load (lbf/ft, lbf/in, k/ft, k/in)
p	depth of fastener penetration into wood members (in)		x	variable distance (ft or in)

symbol	definition (units)
Z	nominal lateral design value for a single fastener connection (lbf)
Z'	allowable lateral design value for a single fastener connection (lbf)
$Z_{m\perp}$	nominal lateral design value for single bolt or lag bolt in wood-to-wood connection with main member loaded perpendicular to grain and side member loaded parallel to grain (lbf)
$Z_{s\perp}$	nominal lateral design value for single bolt or lag bolt in wood-to-wood connection with main member loaded parallel to grain and side member loaded perpendicular to grain (lbf)
$Z_{\parallel}$	nominal lateral design value for a single bolt or lag bolt in connection with all wood members loaded parallel to grain (lbf)
$Z_{\perp}$	nominal lateral design value for a single bolt or lag screw wood-to-wood, wood-to-metal, or wood-to-concrete connection with all wood members loaded perpendicular to grain (lbf)
Δ	actual deflection (in)
Δ_{allow}	maximum allowable deflection (in)
θ	angle between direction of load and direction of wood grain (usually longitudinal axis of member) (degrees)

Structural and Physical Properties of Wood

1. Structure of Wood

The cross section of the tree trunk shown in Fig. 1.1 shows the bark on the outside, the wood on the inside, and the pith at the central core. Wood is composed of thin tubular cells called *fibers*. New wood cells are formed on the inside of new bark cells that are grown under the existing bark. Thus, growth in a tree trunk results from the formation of these new cells.

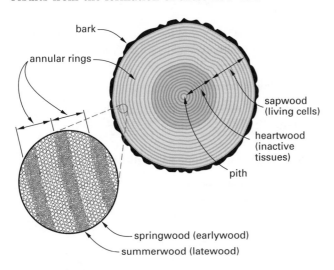

Figure 1.1 Cross Section of Tree Trunk

The wood cell walls are made up of cellulose, and the cells are bound together by a substance known as *lignin*. The wood formed early in the growing season and that formed later are different, producing distinct annual growth rings. Each ring consists of two layers: *springwood* (also called *earlywood*) and *summerwood* (also called *latewood*). The summerwood is relatively heavy, so it has greater density and strength. The wood rings nearest the bark, called *sapwood*, are usually light-colored and the wood inside, called *heartwood*, is relatively darker. There is no difference in mechanical properties between sapwood and heartwood. However, the sapwood is less resistant to decay.

2. Hardwoods and Softwoods

The hardwoods (oak, maple, ash, birch, yellow poplar, etc.) have broad leaves that they lose in the fall or winter. The softwoods (pines, firs, redwoods, etc.) have narrow needle-like leaves that remain on the tree throughout the year; thus, they are evergreen. The terms *hardwood* and *softwood* are unrelated to the actual hardness or strength of the wood.

3. Wood Defects During Growth

A. Knots

Knots are portions of branches that have been enveloped into the trunk of the tree during growth. They reduce the wood strength because they interrupt the fiber directions in a wood member.

B. Reaction Wood

Reaction wood is abnormal wood that forms on the compression or tension side of leaning and crooked trees as a response to the tree's own dead weight. Many properties of reaction wood differ from those of straight wood. Reaction wood should not be used in structural members.

C. Cross Grain

Cross grain is a generic term describing wood fibers (cell walls) that are not aligned with the member's major axis. Cross grain can occur during growth or because of taper cuts for lumber.

D. Shakes

Shakes are cracks that are parallel to the annual growth ring, and they can develop in a standing tree.

4. Sawn Lumber

Wood members that have been manufactured by cutting a piece from a tree log are called *sawn lumber*.

5. Moisture Content

The *moisture content* (MC) of wood varies among species, and even in a single species it depends on locations in the tree trunk. The moisture content in green wood can be anywhere from 30% to more than 200% (based on the oven-dry weight).

A. Fiber Saturation Point

Moisture in green wood that exists in the cell cavities is called *free water*. Moisture that exists in the cell walls is called *bound water*. There is about five times as much free water as there is bound water. The moisture content when the free water has dissipated is called the *fiber saturation point* (FSP). The FSP varies with species, but it is typically about 30%. It is only below the FSP that wood starts to shrink.

B. Equilibrium Moisture Content

Wood continues to dry by losing bound water until the moisture in the wood has come to a balance with that in the surrounding atmosphere. The moisture content at this point is known as the *equilibrium moisture content* (EMC). The EMC in the United States ranges from 5% to 25%, with 10% to 15% the more common range. The EMC can be 5% or lower at a humidity of 30% or less with temperatures greater than 100°F, and it can be 25% or higher at a humidity of 98% with temperatures lower than 100°F (*Wood Handbook*, 1987).

For most buildings, the MC at the time of construction is higher than the EMC, and gradually the MC reaches the EMC during service.

C. Green (Wet), Dry, and Seasoning

The term *green wood* can mean the fresh-cut state of lumber. The terms *green* and *dry* are also used in the National Design Specification (NDS) tables for design values. The term *dry* in the tables refers to areas where moisture content in the wood will not exceed 19% in sawn lumber and 16% in glued laminated timber for an extended time period.

The term *seasoning* usually refers to a controlled drying process by air or kiln drying.

6. Seasoning Defects from Lumber Shrinkage

As the lumber dries below the FSP, shrinkage decreases the size of the cross section and can cause cracks and warping in the lumber (see Figs. 1.2 and 1.3). However, most structural strength properties increase as the MC in the wood decreases to about 15%.

A. Longitudinal Side Checks

The wood near the surface dries faster than the wood at the inner core. This nonuniform drying process causes longitudinal cracks known as *seasoning checks*, *side checks*, or *surface checks* to form near the middle of the wide dimension of lumber lengths.

B. End Checks (Radial Cracks)

The dimensional changes in the wood cross section caused by drying are not uniform. For example, shrinkage parallel to the annual growth ring, called *tangential shrinkage*, is greater than perpendicular shrinkage, called *radial shrinkage*.

Tangential shrinkage causes end checks in the radial direction and is about twice the amount of radial shrinkage.

C. Splits

A *split* is a check occurring across the entire cross section, and it is parallel to the lumber length.

D. Wane

Wane is bark or a lack of wood at the corner(s) of the lumber cross section.

E. Warping

Warping is bowing, crooking, and cupping caused by a difference in tangential and radial shrinkage.

F. Longitudinal Shrinkage

Longitudinal shrinkage is very small and is usually ignored.

G. Shrinkage and Swelling Calculations

The amount of shrinkage or swelling that may occur in a structure in service can be estimated by using Eq. 14-2 in the *Wood Handbook*, 1987.

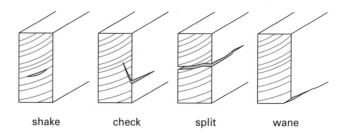

shake   check   split   wane

Figure 1.2 Common Defects in Lumber

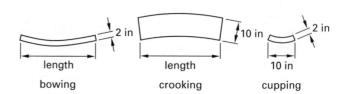

Figure 1.3 Warping for 2 × 10 Lumber

7. Thermal Properties

The thermal conductivity of wood, often given in units of Btu/hr-ft-in-°F, is much less than that of many other materials. For example, it is only about $1/16$ that of sand and gravel, $1/6$ that of clay brick, and $1/400$ that of steel. Therefore, wood is a good insulator.

Since wood is a good insulator, the temperature inside a large timber during a fire is elevated only slightly. Consequently, charred timbers on the surface protect the inside timbers, helping prevent a complete collapse of heavy timber structures.

The thermal coefficient of expansion for wood (parallel to grain) is approximately one-half that of steel. Since the dimensional changes caused by variations in moisture content are much greater than those caused by temperature changes, it is customary to neglect thermal expansion and contraction under most normal conditions.

8. Electrical Properties

Electrical conductivity of wood increases greatly with increases in MC (i.e., resistivity decreases). The electrical resistivity is about 10×10^{12} $\Omega \cdot$m at the oven-dry state and 5×10^3 $\Omega \cdot$m at the FSP (i.e., at about 30% MC). Steel's electrical resistivity is 20×10^{-8} $\Omega \cdot$m.

Mechanical Properties of Lumber and National Design Specifications

The arrangement of fibers (cells) is oriented in concentric cylinders in the tree trunk. As a result, wood has different characteristics in the different directions; that is, wood is anisotropic. For simplification, wood is considered an orthotropic material, requiring consideration in three directions only: longitudinal (parallel to the length of the fibers), radial (perpendicular to the length of the fibers), and tangential (in a direction tangential to the annual growth ring). (See Fig. 2.1.)

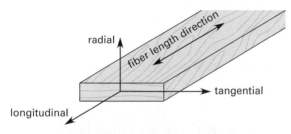

Figure 2.1 Orientation of Axes with Respect to Grain Direction

In order to further simplify these directional options to a practical use, the radial and tangential directions are lumped into a perpendicular-to-grain category. The other category is parallel-to-grain in the longitudinal direction. Strength, modulus of elasticity, and other characteristics, such as shrinkage/swelling and thermal coefficients, differ in the two directions.

1. Lumber Grading

Lumber is sawn from a tree log and is quite variable in mechanical properties. For simplicity and economy for users, lumbers of similar mechanical properties are grouped into a single bracket known as a *grade*. Lumber grade is the classification of lumber with regard to strength and utility in accordance with the grading rules of an approved grading agency.

A. Visual Grading

The majority of sawn lumber is graded by visual inspection in accordance with established grading rules. The grade stamp in lumber includes the commercial grade, the species or species group, and other pertinent information, such as the MC range. For example, S-GRN (surfaced green) means an MC greater than 19% at the time of manufacture, and S-DRY (surfaced dry) means an MC of 19% or less at the time of manufacture.

NDS design values listed in the NDS Supplement are for strength properties of lumber with a moisture content of 19% (16% for glulam) or less for an extended time period.

The design values for most species and grades of visually graded dimension lumber (nominal thickness of 2 in to 4 in) are based on the full-size lumber in-grade test program. Design values of visually graded timbers (5 in minimum nominal dimension) and decking (2 in to 4 in thick and loaded on the wide face) are based on the clear wood testing program. Clear wood strengths are then reduced to account for density, slope of wood grain, knots, shakes, checks, splits, size, test load duration, test moisture content, variability, and so on, to arrive at the NDS table design values.

Lumber grading rules also provide limits on the slope of grain allowed in different stress grades. The slope of grain is measured from a line that is parallel to the longitudinal axis of a lumber length, and it is expressed as a ratio, for example, 1 to 15. Because wood is relatively weaker and more variable in tension perpendicular to grain, designs that include lumber members with steep slopes of grain should be avoided (see Table 2.1).

Table 2.1 Effects of Grain Slopes

slope of grain in member	modulus of rupture	compression parallel to grain
	percentage	
straight-grained	100	100
1 to 25	96	100
1 to 15	89	100
1 to 10	81	99
1 to 5	55	93

Reproduced from *Wood Handbook: Wood as an Engineering Material*, Agricultural Handbook 72, U.S. Government Printing Office, Washington, DC, 1987, Table 4-9.

B. Machine Stress-Rated Grading

Machine stress-rated (MSR) grading (or E-rated grading) is based on a stiffness and deflection relationship. MSR grading machines measure the flatwise bending stiffness in a span of about 4 ft. MSR lumber is stamped with a grade stamp and has less variability in mechanical properties than visually graded lumber. MSR lumber is limited to 2 in or less in thickness. NDS Supp. Table 4C lists design values for MSR lumber.

C. Machine Evaluated Lumber

Machine evaluated lumber (MEL) is nondestructively evaluated by approved mechanical grading equipment. This equipment also evaluates the lumber strength brackets. This lumber has greater variations in F_b/E combinations than MSR lumber. Tabulated design values are given in NDS Supp. Table 4C.

2. National Design Specifications

The 1997 *National Design Specification for Wood Construction* and NDS Supplement (approved August 7, 1997) contain the basic specifications for design values for wood construction. Note that part or all of the NDS is incorporated into major model building codes in the United States.

3. NDS Design Values for Lumber and Glued Laminated Timbers

A. Design Values in the NDS Supplement

Tabulated design values for visually graded structural lumber in the 1997 NDS Supplement have undergone major changes from the earlier NDS editions. New adjustment factors for determining allowable design values from tabulated design values have also been included in the NDS Supplement.

B. Allowable Design Values

Tabulated design values in the NDS Supplement are reduced or increased, usually by applying adjustment factors, to arrive at allowable design values.

$$\text{allowable design value} = (\text{table design value})$$
$$\times (\text{adjustment factors})$$

C. Frequently Used Adjustment Factors

[NDS Table 2.3.1]

Adjustments in design values are for factors including load duration (C_D), wet service condition (C_M), temperature (C_t), beam stability (C_L), size—only to visually graded sawn lumber—(C_F), volume—only to glulam beams—(C_V), flat use (C_{fu}), column stability (C_p), and bearing area (C_b).

Table 2.2 List of Commonly Used Design Properties

design properties	symbol for tabulated design value	symbol for allowable (adjusted) design value
bending stress	F_b	F_b'
tension stress parallel to grain	F_t	F_t'
shear stress parallel to grain	F_v	F_v'
compression stress perpendicular to grain	$F_{c\perp}$	$F_{c\perp}'$
compression stress parallel to grain	F_c	F_c'
modulus of elasticity	E	E'

4. NDS Design Values for Mechanical Connections

[NDS Table 7.3.1 and 7.3.3]

Nominal design values for mechanical connections in the NDS are multiplied by all applicable adjustment factors to obtain the allowable design values.

Adjustment factors for determining allowable design values of mechanical connections include load duration (C_D), wet service condition (C_M), temperature (C_t), group action (C_g), geometry (C_Δ), penetration depth (C_d), end grain (C_{eg}), metal side plate (C_{st}), toe-nail (C_{tn}), and diaphragm (C_{di}).

Lumber Size Categories and Allowable Design Stress

Resistances for wood members are based on various factors, such as species, size, conditions of use, moisture content, load time duration, and so on. These are called *adjustment factors*. Table design values for visually graded and mechanically graded lumber and glulam are specified in NDS Supp. Tables 2A, 4A, 4B, 4C, 4D, 4E, 5A, 5B, and 5C.

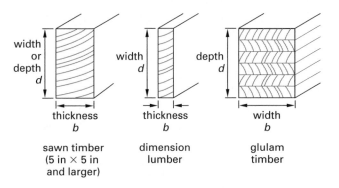

Figure 3.1 Use and Size Categories

1. Size Categories

A. Size Classifications (Sawn Lumber)

Dimension Lumber

Dimension lumber has a nominal thickness of 2 in to 4 in with the load applied to either the narrow or wide face (decking is loaded on the wide face only).

Types of dimension lumber include light framing (width of 2 in to 4 in), structural joists and planks (width of 5 in and wider), studs (width of 2 in to 6 in and lengths limited to a maximum of 10 ft), and decking (width of 4 in or wider and loaded about the minor axis).

Timbers

Timbers have a minimum nominal dimension of 5 in or thicker.

- Beams and stringers refer to a rectangular cross section, nominal 5 in or thicker, with a width of more than 2 in greater than the thickness and loaded on the narrow face during bending. Examples are 6×10, 7×10, and 10×14.

- Posts and timbers refer to square or approximately square cross sections of 5 in $\times$ 5 in nominal dimensions or larger with a width no greater than 2 in more than the thickness. These can be used as posts or columns. Strength in bending is not especially important. The lateral load can be applied to either the narrow face or the wide face. Examples are 6×6, 6×8, 8×10, and 10×12.

- Boards are 1 in to $1^1/2$ in thick, and 2 in or wider.

B. Sizes for Structural Classifications

Lumber dimensions can be specified in one of three different ways: dressed, rough sawn, and full sawn. Dressed lumber has been dressed on a planing machine to obtain smooth surfaces and uniform sizes, and it has actual dimensions that are less than the nominal dimensions usually specified on construction plans. Rough sawn lumber is approximately $1/8$ in larger than dressed sizes (large timbers are commonly rough sawn to the desired dimensions). Full-sawn lumber size is specified by using the actual dimensions and is generally not available (see Table 3.1).

Structural Calculations

Structural calculations are based on net dimensions, not on nominal sizes, for the anticipated use conditions. Shrinkage may have to be accounted for when detailing connections, but standard dimensions are accepted for stress calculations. Dimensions stated in construction plans are nominal dimensions. Net dimensions for dressed lumber thickness are generally taken as $1/2$ in less than nominal. However, the net width of dimension lumber exceeding 6 in in width is taken as $3/4$ in less than nominal.

Table 3.1　Examples of Nominal and Net Dimensions

nominal size b (thickness) × d (width)	dressed size (S4S) $b \times d$
timbers	
6 × 6	$5^{1}/_{2}$ in × $5^{1}/_{2}$ in
8 × 8	$7^{1}/_{2}$ in × $7^{1}/_{2}$ in
dimension lumber	
2 × 6	$1^{1}/_{2}$ in × $5^{1}/_{2}$ in
2 × 8	$1^{1}/_{2}$ in × $7^{1}/_{4}$ in
3 × 6	$2^{1}/_{2}$ in × $5^{1}/_{2}$ in
3 × 10	$2^{1}/_{2}$ in × $9^{1}/_{4}$ in
4 × 6	$3^{1}/_{2}$ in × $5^{1}/_{2}$ in
4 × 8	$3^{1}/_{2}$ in × $7^{1}/_{4}$ in

For 4 in (nominal) and thinner lumber, net *dry* sizes are used in structural calculations regardless of the moisture content at the time of manufacture or use.

For 5 in (nominal) and thicker lumber, net *green* dressed sizes ($^{1}/_{2}$ in less than nominal) are used in structural calculations regardless of the moisture content at the time of manufacture or use.

C. Glued Laminated Timber

Structural calculations are either based on the net finished dimensions as specified in NDS Supp. Table 1C or are in accordance with the manufacturer's data.

2. Table Design Values Specified

[NDS Supp. Tables 4 and 5]

A. Visually Graded and Mechanically Graded Lumber

Table design values for visually graded and mechanically graded sawn lumber given in NDS Supp. Tables are for a 10-year load duration and dry service conditions (moisture content less than or equal to 19%).

Table design values for bending in sawn lumber apply to either the x-axis or y-axis for dimension lumber, and to posts and timbers. For beams and stringers, the design values apply to x-axis bending only (load on the narrow face, b), and for decking the design values apply to y-axis bending (load on the wide face, d).

B. Glued Laminated Timber

Table design values for glued laminated timber are for a 10-year load duration and for dry service conditions (moisture content less than or equal to 16%). The design values for bending, F_b, apply to a glulam beam with width $b = 5^{1}/_{8}$ in, depth $d = 12$ in, and length $L = 21$ ft. Therefore, the volume-effect factor, C_V, is used to obtain the values for other sizes of glulams.

3. Adjustment of Tabulated Design Values

[NDS 2.3 and Table 2.3.1; NDS 4.3; NDS Supp. Tables 4 and 5 Adj. Factors]

NDS Supp. Table design values (F_b, F_t, F_v, $F_{c\perp}$, F_c, E, F_g, etc.) are multiplied by all applicable adjustment factors to determine allowable design values (F_b', F_t', F_v', $F_{c\perp}'$, F_c', E', F_g', etc.) for actual conditions of use in accordance with the NDS cited above. Note: The footnotes in the NDS Tables are important and should not be ignored.

$$\text{allowable design value} = (\text{table design value})$$
$$\times (\text{adjustment factors})$$

4. Adjustment Factors

[NDS 2.3, 4.3, and NDS Supp. Tables 4 and 5]

NDS Tables 2.3.1 and 2.3.2 are duplicated in Tables 3.2 and 3.3. Some of the footnotes for these tables will be explained in later design sections.

Following are some of the frequently used adjustment factors.

A. Load Duration Factor, C_D

[NDS Table 2.3.2 and App. B]

Wood can sustain greater maximum loads for short-load durations than for long-load durations. When the full maximum load is applied either cumulatively or continuously for periods fewer than 10 years, NDS Table design stresses as well as mechanical fastenings are multiplied by the load duration factor, C_D.

B. Wet Service Factor, C_M

[NDS 2.3.3; NDS Supp. Tables 4A, 4B, 4C, 4D, 4E, 5A, 5B, and 5C]

When the moisture content at installation or in service is expected to exceed 19% for sawn lumber and 16% for glulam, table design values are reduced by the wet service factors, C_M, given in NDS Supp. Tables 4 and 5.

C. Size Factor, C_F

[NDS 4.3.2; NDS Supp. Table 4 Adj. Factors]

The size factor, C_F, for dimension lumber (i.e., 2 in to 4 in thick) for bending, tension, and compression parallel to grain is specified in NDS Supp. Tables 4A and 4B Adj. Factors.

For timbers (i.e., 5 in × 5 in and larger with the depth exceeding 12 in), C_F applies for bending only and is given in NDS Supp. Table 4D Adj. Factors as

$$C_F = \left(\frac{12}{d}\right)^{1/9} \leq 1.0$$

Table 3.2 Applicability of Adjustment Factors (NDS Table 2.3.1)

	load duration factor	wet service factor	temperature factor	beam stability factor [a]	size factor [b]	volume factor [a,c]	flat use factor [a,c]	incising factor [d]	repetitive member factor [e]	curvature factor [f]	form factor	column stability factor	shear stress factor [g]	buckling stiffness factor [h]	bearing area factor
$F_b' = F_b$	C_D	C_M	C_t	C_L	C_F	C_V	C_{fu}	C_i	C_r	C_c	C_f	–	–	–	–
$F_t' = F_t$	C_D	C_M	C_t	–	C_F	–	–	C_i	–	–	–	–	–	–	–
$F_v' = F_v$	C_D	C_M	C_t	–	–	–	–	C_i	–	–	–	–	C_H	–	–
$F_{c\perp}' = F_{c\perp}$	–	C_M	C_t	–	–	–	–	C_i	–	–	–	–	–	–	C_b
$F_c' = F_c$	C_D	C_M	C_t	–	C_F	–	–	C_i	–	–	–	C_P	–	–	–
$E' = E$	–	C_M	C_t	–	–	–	–	C_i	–	–	–	–	–	C_T	–
$F_g' = F_g$	C_D	–	C_t	–	–	–	–	–	–	–	–	–	–	–	–

[a]The beam stability factor, C_L, shall not apply simultaneously with the volume factor, C_V, for glued laminated timber bending members (NDS 5.3.2). Therefore the lesser of these adjustment factors shall apply.

[b]The size factor, C_F, shall apply only to visually graded sawn lumber members and to round timber bending members (NDS 4.3.2).

[c]The volume factor, C_V, shall apply only to glued laminated timber bending members (NDS 5.3.2).

[d]The flat use factor, C_{fu}, shall apply only to dimension lumber bending members 2 in to 4 in (nominal) thick (NDS 4.3.3) and to glued laminated timber bending members (NDS 5.3.3).

[e]The repetitive member factor, C_r, shall apply only to dimension lumber bending members 2 in to 4 in thick (NDS 4.3.4).

[f]The curvature factor, C_c, shall apply only to curved portions of glued laminated timber bending members (NDS 5.3.4).

[g]Shear design values parallel to grain, F_v, for sawn lumber members shall be permitted to be multiplied by the shear stress factors, C_H, specified in Tables 4A, 4B, 4C and 4D.

[h]The buckling stiffness factor, C_T, shall apply only to 2 in × 4 in or smaller sawn lumber truss compression chords subjected to combined flexure and axial compression when 3/8 in or thicker plywood sheathing is nailed to the narrow face (NDS 4.4.3).

Reproduced from *National Design Specification for Wood Construction*, 1997 Edition, courtesy of American Forest & Paper Association, Washington, D.C.

Bending design values for decking are based on 4 in thick decking. For all species of 2 in thick or 3 in thick decking except redwood, C_F for bending is given in NDS Supp. Table 4E, Adj. Factors.

Table 3.3 Frequently Used Load Duration Factors, C_D^a (NDS Table 2.3.2)

load duration	C_D	typical design loads
permanent	0.9	dead load
ten years	1.0	occupancy live load
two months	1.15	snow load
seven days	1.25	construction load
ten minutes	1.6	wind/earthquake load
impact[b]	2.0	impact load

[a]Load duration factors shall not apply to modulus of elasticity, E, nor to compression perpendicular to grain design values, $F_{c\perp}$, based on a deformation limit.

[b]The impact load duration factor shall not apply to structural members pressure-treated with water-borne preservatives to the heavy retentions required for "marine" exposure (see *AWPA Book of Standards*, American Wood Preservers' Association, Stevensville, MD, 1991), nor to structural members pressure-treated with fire-retardant chemicals. The impact load duration factor shall not apply to connections.

Reproduced from *National Design Specification for Wood Construction*, 1997 Edition, courtesy of American Forest & Paper Association, Washington, D.C.

D. Flat Use Factor, C_{fu}

[NDS 4.3.3; NDS Supp. Tables 4 and 5 Adj. Factors]

When dimension lumber (i.e., 2 in to 4 in thick) is loaded in bending on the wide face, C_{fu} is given in NDS Supp. Tables 4A, 4B, and 4C, Adj. Factors.

Tabulated bending design values for decking have already been adjusted for flatwise usage (load applied to wide face) in NDS Supp. Table 4E.

E. Form Factor, C_f

[NDS 2.3.8]

These form factors insure that a circular or diamond-shaped bending member has the same moment capacity as a square bending member having the same cross-sectional area. For a round section, $C_f = 1.18$, and for a diamond section, $C_f = 1.414$.

F. Bearing Area Factor, C_b

[NDS 2.3.10]

Tabulated compression design values perpendicular to grain, $F_{c\perp}$, apply to bearings of any length at the ends of a member and to all bearings 6 in or more in length at any other location. For bearings less than 6 in in length and not nearer than 3 in to the end of a member, see NDS Table 2.3.10.

G. Incising Factor for Structural Sawn Lumber, C_i

[NDS 2.3.11]

Tabulated design values shall be multiplied by an incising factor, C_i (NDS Table 2.3.11), when structural sawn lumber is incised to increase penetration of preservatives. Incisions are cut parallel to grain a maximum depth of 3/4 in, a maximum length of 3/8 in, and a maximum density of 357 incisions per square foot.

Other adjustment factors, such as temperature factor, C_t (equals 1.0 for 100°F or less), preservative treatment, and others are described in NDS 2.3 and in other relevant sections of the NDS. (Other adjustment factors will be covered as they are needed.)

Example 3.1
Allowable Design Values: Dimension Lumber

A 2×8 no. 1 douglas fir-larch is loaded on its narrow face. It is adequately braced to prevent lateral torsional buckling. The loads consist of dead load plus snow load (DL + SL). The wet service condition prevails (moisture content greater than 19%). Assume normal temperature.

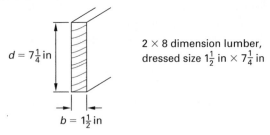

$d = 7\frac{1}{4}$ in

2×8 dimension lumber, dressed size $1\frac{1}{2}$ in $\times 7\frac{1}{4}$ in

$b = 1\frac{1}{2}$ in

Find the relevant allowable design values in accordance with NDS specifications.

Solution:

NDS Section

2×8 is a visually graded dimension lumber.

Table 2.3.2   For snow load, $C_D = 1.15$.

Table 2.3.1;
2.3; Supp.
Table 4A
and 4A Adj.
Factors

$$F'_b = F_b C_D C_M C_t C_L C_F C_V C_{fu} C_r C_c C_f$$
$$= \left(1000 \ \frac{\text{lbf}}{\text{in}^2}\right)(1.15)(0.85)(1.0)(1.0)$$
$$\times (1.2)(1.0)(1.0)(1.0)(1.0)(1.0)$$
$$= 1173.0 \ \text{lbf/in}^2$$

$$F'_t = F_t C_D C_M C_t C_F$$
$$= \left(675 \ \frac{\text{lbf}}{\text{in}^2}\right)(1.15)(1.0)(1.0)(1.2)$$
$$= 931.5 \ \text{lbf/in}^2$$

$$F'_v = F_v C_D C_M C_t C_H$$
$$= \left(95 \ \frac{\text{lbf}}{\text{in}^2}\right)(1.15)(0.97)(1.0)(1.0)$$
$$\text{[assume splits]}$$
$$= 106.0 \ \text{lbf/in}^2$$

$$E' = E C_M C_t$$
$$= \left(1,700,000 \ \frac{\text{lbf}}{\text{in}^2}\right)(0.9)(1.0)$$
$$= 1,530,000 \ \text{lbf/in}^2$$

Supp.
Table 2A

$$F'_g = F_g C_D C_t$$
$$= \left(1350 \ \frac{\text{lbf}}{\text{in}^2}\right)(1.15)(1.0)$$
$$= 1552.5 \ \text{lbf/in}^2$$

Example 3.2
Allowable Design Values: Beams and Stringers

A full-sawn 6×10 no. 2 red oak in bending is subjected to a combination loading of dead load, live load, snow load, and wind load (DL + LL + SL + WL). Assume use conditions to be dry service, normal temperature, adequate lateral bracing, and splits on the wide face of the 6×10.

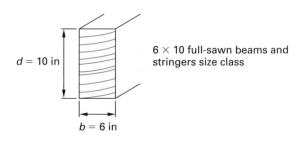

$d = 10$ in

6×10 full-sawn beams and stringers size class

$b = 6$ in

Find the relevant allowable design values in accordance with NDS specifications.

Solution:

NDS Section

The size classification is
thickness = 6 in > 5 in
width = 10 in > (thickness + 2 in)

Therefore, the classification is beams and stringers, visually graded.

Table 2.3.2   For wind load, $C_D = 1.6$.

Table 2.3.1;
Supp.
Table 4D
and 4D Adj.
Factors

$$F'_b = F_b C_D C_M C_t C_L C_F$$
$$= \left(725 \ \frac{\text{lbf}}{\text{in}^2}\right)(1.6)(1.0)(1.0)(1.0)(1.0)$$
$$= 1183.2 \ \text{lbf/in}^2$$

$$F'_t = F_t C_D C_M C_t$$
$$= \left(375 \ \frac{\text{lbf}}{\text{in}^2}\right)(1.6)(1.0)(1.0)$$
$$= 600.0 \ \text{lbf/in}^2$$

(C_F is not applicable to tension in timbers, i.e., 5×5 and larger.)

$$F'_v = F_v C_D C_M C_t C_H$$
$$= \left(80 \ \frac{\text{lbf}}{\text{in}^2}\right)(1.6)(1.0)(1.0)(1.0)$$
$$= 128.0 \ \text{lbf/in}^2$$

$$F'_{c\perp} = F_{c\perp} C_M C_t C_b$$
$$= \left(820 \ \frac{\text{lbf}}{\text{in}^2}\right)(1.0)(1.0)(1.0)$$
$$= 820.0 \ \text{lbf/in}^2$$

Beam Design: Sawn Lumber

The design of rectangular sawn wood beams includes bending (including lateral stability), shear, deflection, and bearing at supports and load points.

1. Bending

[NDS 3.2–3.5; 2.3.1]

Allowable bending design values, F_b', will be greater than the actual (calculated) bending stress, f_b, given by

$$f_b = \frac{Mc}{I} = \frac{M}{S} \leq F_b' \qquad \text{[NDS Eq. 3.3-1]}$$

M = actual (calculated) moment

$$S = \frac{I}{c}$$

A. Section Modulus

[NDS 3.2.1]

The section modulus is for the net, dressed section. The span for calculating bending moment, M, is taken as the distance from face to face, plus one-half the required bearing length at each end.

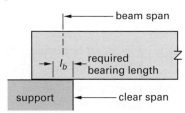

Figure 4.1 Beam Span

B. Allowable Bending Design Values

[NDS Supp. Table 4; NDS Table 2.3.1]

The allowable bending stresses for sawn lumber beams of rectangular cross section are summarized here. Some obvious adjustment factors can default to 1.0 and can be used to obtain allowable bending stresses from the NDS Table values, as listed in NDS Table 2.3.1. As a result, some of these factors are often omitted. For example, for covered structures with a normal temperature, C_M for the wet service factor and C_t for the temperature factor are 1.0 and do not have to be included in the calculations.

Figure 4.2 Sawn Lumber Beam Bending About the Strong Axis

The allowable bending stress for the strong axis (x-axis) is given by

$$F_{bx}' = F_{bx} C_D C_M C_t C_L C_F C_r$$

$F_{bx} = F_b$ = tabulated bending stress. Recall that for sawn lumber, tabulated values of bending stress apply to the x-axis (except decking). Values are listed in NDS Supp. Tables 4A and 4B for dimension lumber and in Table 4D for timbers.

C_D (See NDS Table 2.3.2; NDS App. B.)

$C_M = 1.0$ for $M_C \leq 19\%$ (as in most covered structures) (See NDS Supp. Tables 4A, 4B, and 4D.)

$C_t = 1.0$ for normal temperature conditions (See NDS Table 2.3.4.)

$C_L = 1.0$ for continuous lateral support of the compression face of the beam. For other conditions, compute C_L in accordance with NDS 3.3.3.

C_F = Obtain the values from the Adj. Factors section of NDS Supp. Tables 4A and 4B for dimension lumber and in Table 4D for timbers. (See NDS 4.3.2.)

$C_r = 1.15$ for dimension lumber applications that meet the definition of a repetitive member (See NDS 4.3.4 and also NDS Supplement Tables 4A, 4B, and 4C, Adjustment Factors.)
$= 1.0$ for all other conditions.

C_i = incising factor (NDS 2.3.11)
$= 1.0$ for all other conditions

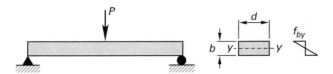

Figure 4.3 Sawn Lumber Beam Bending About the Weak Axis

The allowable bending stress for the weak axis (y-axis) is given by

$$F'_{by} = F_{bx}C_D C_M C_t C_F C_{fu} C_i$$

$F_{bx} = F_b =$ bending stress given in NDS Tables, which are based on edgewise use (load applied to narrow face). Values of F_b are listed in NDS Supp. Tables 4A and 4B for dimension lumber and in Table 4D for timbers. Recall that the load can also be applied to the y-axis for all sizes of sawn lumber except beams and stringers.

$C_{fu} =$ Obtain values from the Adj. Factors section of NDS Supp. Tables 4A, 4B, and 4C for dimension lumber loaded on the wide face. Contact the appropriate lumber rules-writing agency to obtain C_{fu} for beam and stringer values. (See NDS 4.3.3.)

$= 1.0$ for posts and timbers; that is, F_b for posts and timbers in NDS Table 4D are for either x-axis or y-axis bending.

The form factor, C_f, must also be included for beams that have nonrectangular cross sections (see NDS 2.3.8).

C. Load Duration Factor, C_D

[NDS 2.3.2; NDS App. B]

The term *duration of load* refers to the total accumulated length of time that a load is applied during the life of a structure. Furthermore, in considering the duration, it is the full design load that is applied, not the length of time over which a portion of the load is applied. The factor, C_D, is associated with the shortest duration load in a given load combination. When both wind and earthquake loads are possible, there is no need to assume that they act simultaneously for normal design situations.

D. Beam Stability Factor, C_L

[NDS 3.3.3 and 4.4.1]

If the compression zone of the beam is not braced to prevent lateral torsional buckling, the beam may buckle at a bending stress that is less than the allowable design stress when buckling is prevented.

Rule of Thumb Method

[NDS 4.4.1]

C_L is 1.0 if d/b, based on nominal dimensions, is

- 2; no lateral support is required.
- 3 or 4; the beam ends are held in position.
- 5; the compression edge of a beam is laterally supported throughout its length.
- 6; bridging or full-depth solid blocking is provided at 8 ft intervals or less.
- 7; both edges are held in line for their entire length.

Analytic Method (when $d > b$)

[NDS 3.3.3.4; NDS Table 3.3.3]

- Lateral support will be provided at points of bearing to prevent rotation and/or lateral displacement at bearing points.
- The unsupported beam length, l_u, is the distance between points of end bearings or between points of intermediate lateral support preventing rotation and/or lateral displacement.

The beam stability factor is given by

$$C_L = \frac{1 + \dfrac{F_{bE}}{F^*_{bx}}}{1.9} - \sqrt{\left(\frac{1 + \dfrac{F_{bE}}{F^*_{bx}}}{1.9}\right)^2 - \frac{\dfrac{F_{bE}}{F^*_{bx}}}{0.95}}$$ [NDS Eq. 3.3-6]

The critical buckling (Euler) value for bending member is

$$F_{bE} = \frac{K_{bE}E'_y}{R_B^2}$$

The tabulated bending stress for x-axis multiplied by certain adjustment factors is

$$F^*_{bx} = F_{bx}\left(\begin{array}{c}\text{product of all adjustment factors}\\ \text{except } C_{fu}, \ C_V, \text{ and } C_L\end{array}\right)$$

$K_{bE} = 0.438$ for visually graded lumber and MEL

$= 0.609$ for glulam and MSR lumber
[$\text{COV}_E \le 0.11$ (See NDS App. F.2.)]

$E'_y =$ modulus of elasticity associated with lateral torsional buckling

$=$ modulus of elasticity about the y-axis multiplied by all appropriate adjustment factors

$= E_y C_M C_t$
Recall that C_D does not apply to E. For sawn lumber, $E_y = E_x$. For glulam, E_x and E_y may be different.

The slenderness ratio of unbraced beam length is

$$R_B = \sqrt{\frac{l_e d}{b^2}}$$

l_e (See Table 4.1 or NDS Table 3.3.3.)

Table 4.1 Effective Unbraced Length, l_e, for Bending Members (NDS Table 3.3.3)

cantilever[a]	when $l_u/d < 7$	when $l_u/d \geq 7$
uniformly distributed load	$l_e = 1.33l_u$	$l_e = 0.90l_u + 3d$
concentrated load at unsupported end	$l_e = 1.87l_u$	$l_e = 1.44l_u + 3d$

single-span beam[a]	when $l_u/d < 7$	when $l_u/d \geq 7$
uniformly distributed load	$l_e = 2.06l_u$	$l_e = 1.63l_u + 3d$
concentrated load at center with no intermediate lateral support	$l_e = 1.80l_u$	$l_e = 1.37l_u + 3d$
concentrated load at center with lateral support at center	$l_e = 1.11l_u$	
two equal concentrated loads at one-third points with lateral support at one-third points	$l_e = 1.68l_u$	
three equal concentrated loads at one-fourth points with lateral support at one-fourth points	$l_e = 1.54l_u$	
four equal concentrated loads at one-fifth points with lateral support at one-fifth points	$l_e = 1.68l_u$	
five equal concentrated loads at one-sixth points with lateral support at one-sixth points	$l_e = 1.73l_u$	
six equal concentrated loads at one-seventh points with lateral support at one-seventh points	$l_e = 1.78l_u$	
seven or more equal concentrated loads, evenly spaced, with lateral support at points of load application	$l_e = 1.84l_u$	
equal end moments	$l_e = 1.84l_u$	

[a]For single-span or cantilever bending members with loading conditions not specified in Table 4.1:

$l_e = 2.06l_u$ when $l_u/d < 7$

$l_e = 1.63l_u + 3d$ when $7 \leq l_u/d \leq 14.3$

$l_e = 1.84l_u$ when $l_u/d > 14.3$

Reproduced from *National Design Specification for Wood Construction*, 1997 Edition, courtesy of American Forest & Paper Association, Washington, D.C.

2. Shear Parallel to Grain (Horizontal Shear)

[NDS 3.4; 4.4.2]

The actual shear stress parallel to grain, f_v, shall not be greater than the allowable shear stress, F_v'.

$$f_v = \frac{VQ}{Ib} = \frac{3V}{2bd} \quad \text{[for a rectangular cross section]}$$

$$F_v' = F_v C_D C_M C_t C_i C_H$$

F_v (See NDS Supp. Tables 4A, 4B, and 4D.)

$C_H = 1.0$ conservative for sawn lumber

$ = 1.0$ always for glulam

(See NDS Supp. Tables 4A, 4B, and 4D.)

When calculating the maximum shear force, V, in bending members, the following rules apply (NDS 3.4.3).

a) All loads within a distance from the face of supports equal to the depth, d, of the bending member are ignored.

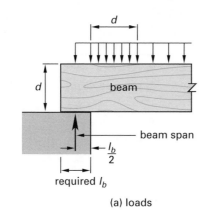

(a) loads

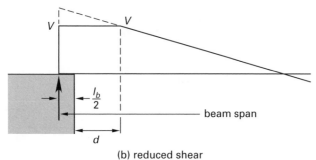

(b) reduced shear

Figure 4.4 Shear at Supports

b) When there is a single moving load or one moving load that is considerably greater than the others, place that load at a distance from each support equal to d; place any other loads in their normal relation.

c) When there are two or more moving loads of about equal weight and in proximity, place the loads in the position that produces the maximum shear force, V, neglecting any loads within a distance d from each support. Loads within d are ignored for shear force determination.

3. Deflection

[NDS 3.5]

Deflections are determined in the same way as with other materials.

$$\Delta = f\left[\frac{P, w, L}{E'I}\right]$$

$E' = EC_M C_t C_T C_i$ (See NDS Table 2.3.1.)

$C_T = 1.0$ for beams

= other than 1.0 for a 2 in × 4 in or smaller sawn lumber truss compression chord when subjected to combined flexure and axial compression (See NDS 4.4.3.)

For long-term loading (NDS 3.5.2), deflection shall be calculated with a creep factor of 1.5 (for unseasoned lumber, 2.0).

4. Bearing

[NDS 2.3.10; 3.10]

A. Bearing Perpendicular to Grain

Bearing stress perpendicular to the grain of wood occurs at beam supports or where loads from one wood member are transferred to another member.

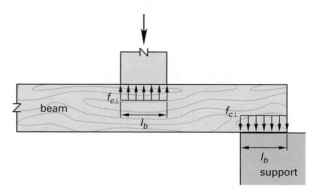

Figure 4.5 Compression (Bearing) Perpendicular to Grain

The actual compressive stress perpendicular to grain will not exceed the allowable compressive design value perpendicular to grain.

$$f_{c\perp} = \frac{P}{A} \leq F'_{c\perp}$$

The allowable compressive (bearing) design value is given by

$$F'_{c\perp} = F_{c\perp} C_M C_t C_i C_b$$

$F_{c\perp}$ = tabulated design value for compression perpendicular to grain, applicable to bearing 6 in or more in length and not nearer than 3 in to the end of a beam (See NDS Supp. Tables 4A, 4B, and 4D.)

This bearing area is

$A = l_b b$

$C_b = \dfrac{l_b + \frac{3}{8}\text{ in}}{l_b}$ $\quad\left[\begin{array}{l}\text{for } l_b \text{ less than 6 in but at}\\ \text{least 3 in from a member end}\end{array}\right]$

= 1.0 for l_b greater than 6 in (see NDS 2.3.10) and at least 3 in from a member end

B. Bearing Parallel to Grain

[NDS Supp. Table 2A]

This type of bearing stress occurs when two wood members bear end to end with each other as well as when a wood end bears on other surfaces.

The actual bearing stress shall not exceed the allowable design value given by

$$f_g = \frac{P}{A} \leq F'_g$$

The allowable compressive (bearing) stress parallel to the grain is

$$F'_g = F_g C_D C_t$$

$$F_g \quad \text{(See NDS Supp. Table 2A.)}$$

C. Bearing Stress Occurring at an Angle Other than 0° or 90° with Respect to Grain Direction

[NDS 3.10.3]

Bearing at some angle to grain other than 0° or 90° is checked by the Hankinson formula.

$$f_\theta = \frac{P}{A} \le F'_\theta = \frac{F'_g F'_{c\perp}}{F'_g \sin^2 \theta + F'_{c\perp} \cos^2 \theta} \quad \text{[NDS Eq. 3.10-1]}$$

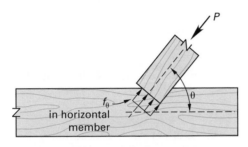

Figure 4.6 Bearing at an Angle to Grain

Example 4.1
Sawn Lumber Beam Analysis

The select structural southern pine 4×16 beam of 20 ft span supports a hoist located at the center of the span. Assume normal load duration (i.e., 10 yr) and dry service conditions. Assume wood density is 37.5 lbf/ft³. Lateral support is provided only at the beam ends. The hoist load, P, is 3000 lbf. The bearing length at each beam support is 4 in. The allowable live load deflection criterion is $L/240$, where L is the beam span length.

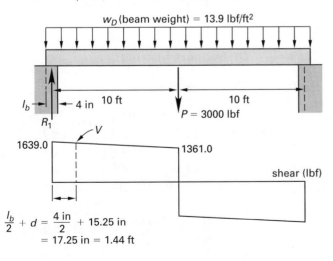

$$\frac{l_b}{2} + d = \frac{4 \text{ in}}{2} + 15.25 \text{ in}$$
$$= 17.25 \text{ in} = 1.44 \text{ ft}$$

Review the beam.

Solution:

NDS Section

For 4×16 (3.5 in $\times$ 15.25 in actual) dimension lumber,

$$I_x = 1034.4 \text{ in}^4$$

$$A = 53.4 \text{ in}^2$$

$$S_x = 135.7 \text{ in}^3$$

$$\begin{aligned} w_D &= \rho A \\ &= \left(37.5 \, \frac{\text{lbf}}{\text{ft}^3}\right)(53.4 \text{ in}^2)\left(\frac{1 \text{ ft}^2}{144 \text{ in}^2}\right) \\ &= 13.9 \text{ lbf/ft} \end{aligned}$$

A. Bending

Table 2.3.1
Supp.

$$F'_b = F_b C_D C_M C_t C_L C_F C_i C_r = F^*_b C_L$$

Table 4B
Adj. Factors

$$\begin{aligned} F^*_b &= F_b C_D C_M C_t C_F C_i C_r \\ &= \left(1900 \, \frac{\text{lbf}}{\text{in}^2}\right)(1.0)(1.0)(1.0) \\ &\quad \times (0.9)(1.0)(1.0) \\ &= 1710 \text{ lbf/in}^2 \end{aligned}$$

$$\begin{aligned} E' &= E'_y = E'_x = E C_M C_t C_i \\ &= \left(1{,}800{,}000 \, \frac{\text{lbf}}{\text{in}^2}\right)(1.0)(1.0)(1.0) \\ &= 1{,}800{,}000 \text{ lbf/in}^2 \end{aligned}$$

3.3.3
Eq. 3.3-6

Find the beam stability factor, C_L.

The unbraced length is

$$l_u = (20 \text{ ft})\left(12 \, \frac{\text{in}}{\text{ft}}\right) = 240 \text{ in}$$

$$\frac{l_u}{d} = \frac{240 \text{ in}}{15.25 \text{ in}} = 15.7 > 7$$

Table 3.3.3

The effective unbraced length is
$$\begin{aligned} l_e &= 1.37 l_u + 3_d \\ &= (1.37)(240 \text{ in}) + (3)(15.25 \text{ in}) = 374.6 \text{ in} \end{aligned}$$

$$\begin{aligned} R_B &= \sqrt{\frac{l_e d}{b^2}} = \sqrt{\frac{(374.6 \text{ in})(15.25 \text{ in})}{(3.5 \text{ in})^2}} \\ &= 21.6 \end{aligned}$$

3.3.3.8

$$K_{bE} = 0.438$$

$$\begin{aligned} F_{bE} &= \frac{K_{bE} E'_y}{R_B^2} = \frac{(0.438)\left(1{,}800{,}000 \, \frac{\text{lbf}}{\text{in}^2}\right)}{(21.6)^2} \\ &= 1690 \text{ lbf/in}^2 \end{aligned}$$

$$C_L = \frac{1 + \dfrac{F_{bE}}{F_b^*}}{1.9} - \sqrt{\left(\frac{1 + \dfrac{F_{bE}}{F_b^*}}{1.9}\right)^2 - \frac{\dfrac{F_{bE}}{F_b^*}}{0.95}}$$

$$= \frac{1 + \dfrac{1690 \; \dfrac{\text{lbf}}{\text{in}^2}}{1710 \; \dfrac{\text{lbf}}{\text{in}^2}}}{1.9}$$

$$- \sqrt{\left(\frac{1 + \dfrac{1690 \; \frac{\text{lbf}}{\text{in}^2}}{1710 \; \frac{\text{lbf}}{\text{in}^2}}}{1.9}\right)^2 - \frac{\dfrac{1690 \; \frac{\text{lbf}}{\text{in}^2}}{1710 \; \frac{\text{lbf}}{\text{in}^2}}}{0.95}}$$

$$= 0.812$$

$$F_b' = F_b^* C_L = \left(1710 \; \frac{\text{lbf}}{\text{in}^2}\right)(0.812)$$

$$= 1388.52 \; \text{lbf/in}^2$$

$$M = \frac{w_D l^2}{8} + \frac{Pl}{4}$$

$$= \frac{\left(13.9 \; \dfrac{\text{lbf}}{\text{ft}}\right)(20 \; \text{ft})^2 \left(12 \; \dfrac{\text{in}}{\text{ft}}\right)}{8}$$

$$+ \frac{(3000 \; \text{lbf})(20 \; \text{ft})\left(12 \; \dfrac{\text{in}}{\text{ft}}\right)}{4}$$

$$= 188{,}340 \; \text{in-lbf}$$

$$f_b = \frac{M}{S_x} = \frac{188{,}340 \; \text{in-lbf}}{135.7 \; \text{in}^3} = 1387.9 \; \text{lbf/in}^2$$

$$F_b' = 1388.52 \; \frac{\text{lbf}}{\text{in}^2} > f_b = 1387.9 \; \text{lbf/in}^2 \; [\text{OK}]$$

B. Shear

NDS Table 2.3.1

$$F_v' = F_v C_D C_M C_t C_i C_H$$

$$= \left(90 \; \frac{\text{lbf}}{\text{in}^2}\right)(1.0)(1.0)(1.0)(1.0)(1.0)$$

$$= 90.0 \; \text{lbf/in}^2$$

The reaction is

$$R_1 = \frac{w_D L}{2} + \frac{P}{2}$$

$$= \left(13.9 \; \frac{\text{lbf}}{\text{ft}}\right)(20 \; \text{ft})\left(\frac{1}{2}\right)$$

$$+ (3000 \; \text{lbf})\left(\frac{1}{2}\right)$$

$$= 1639.0 \; \text{lbf}$$

NDS 3.4.2; 3.4.3

$$V = 1639.0 \; \text{lbf} - \left(13.9 \; \frac{\text{lbf}}{\text{ft}}\right)(1.44 \; \text{ft})$$

$$= 1619 \; \text{lbf}$$

$$f_v = (1.5)\left(\frac{V}{bd}\right) = (1.5)\left(\frac{1619 \; \text{lbf}}{53.38 \; \text{in}^2}\right)$$

$$= 45.5 \; \text{lbf/in}^2$$

$$F_v' = 90 \; \frac{\text{lbf}}{\text{in}^2} > f_v = 45.5 \; \text{lbf/in}^2 \; [\text{OK}]$$

C. Deflection (live load only)

$$\Delta = \frac{Pl^3}{48 E' I}$$

$$= \frac{(3000 \; \text{lbf})\left((20 \; \text{ft})\left(12 \; \frac{\text{in}}{\text{ft}}\right)\right)^3}{(48)\left(1{,}800{,}000 \; \frac{\text{lbf}}{\text{in}^2}\right)(1034.4 \; \text{in}^4)}$$

$$= 0.464 \; \text{in}$$

$$\frac{L}{240} = \frac{(20 \; \text{ft})\left(12 \; \frac{\text{in}}{\text{ft}}\right)}{240}$$

$$= 1.0 \; \text{in} > 0.464 \; \text{in} \quad [\text{OK}]$$

2.3.10; 3.10

D. Support bearing

The support reaction is

$$\frac{w_D L}{2} + \frac{P}{2} = \left(13.9 \; \frac{\text{lbf}}{\text{ft}}\right)(20 \; \text{ft})\left(\frac{1}{2}\right)$$

$$+ \frac{3000 \; \text{lbf}}{2}$$

$$= 1639.0 \; \text{lbf}$$

$$f_{c\perp} = \frac{R_1}{l_b b} = \frac{1639.0 \; \text{lbf}}{(4 \; \text{in})(3.5 \; \text{in})} = 117.1 \; \text{lbf/in}^2$$

$$C_b = 1.0$$

Supp.
Table 4B

$$F'_{c\perp} = F_{c\perp}C_M C_t C_i C_b$$
$$= \left(565 \, \frac{\text{lbf}}{\text{in}^2}\right)(1.0)(1.0)(1.0)(1.0)$$
$$= 565 \text{ lbf/in}^2$$

$$F'_{c\perp} = 565 \, \frac{\text{lbf}}{\text{in}^2} > f_{c\perp} = 117.1 \text{ lbf/in}^2 \text{ [OK]}$$

Example 4.2
Bearing at an Angle to Grain

Assume no. 2 douglas fir-larch lumber, and that C_D, C_M, and C_t are 1.0.

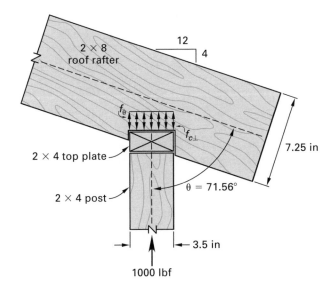

Review the bearing stresses for a roof rafter/post joint.

Solution:

NDS Section

Supp.
Table 4A

$$F_{c\perp} = 625 \text{ lbf/in}^2 \quad [C_F \text{ not applicable}]$$

Supp.
Table 2A Dry

$$F_g = 2020 \text{ lbf/in}^2$$

A. 2 × 4 top plate (with 2 × 4 post)

2.3.10;
Table 2.3.1

Since $l_b = 1.5 \text{ in} < 6 \text{ in}$,

$$C_b = \frac{l_b + \frac{3}{8} \text{ in}}{l_b} = \frac{1.5 \text{ in} + \frac{3}{8} \text{ in}}{1.5 \text{ in}} = 1.25$$

$$F'_{c\perp} = F_{c\perp}C_M C_t C_b$$
$$= \left(625 \, \frac{\text{lbf}}{\text{in}^2}\right)(1.0)(1.0)(1.25)$$
$$= 781.2 \text{ lbf/in}^2$$

$$f_{c\perp} = \frac{P}{A} = \frac{P}{l_b b} = \frac{1000 \text{ lbf}}{(1.5 \text{ in})(3.5 \text{ in})}$$
$$= 190.5 \text{ lbf/in}^2$$

$$F'_{c\perp} = 781.2 \, \frac{\text{lbf}}{\text{in}^2} > f_{c\perp} = 190.5 \text{ lbf/in}^2$$
$$781.2 \text{ lbf/in}^2 > 190.5 \text{ lbf/in}^2$$
$$F'_{c\perp} > f_{c\perp} \quad \text{[OK]}$$

B. 2 × 8 rafter (with 2 × 4 top plate)

$$f_\theta = \frac{P}{A} = \frac{P}{\text{contact area}} = \frac{1000 \text{ lbf}}{(3.5 \text{ in})(1.5 \text{ in})}$$
$$= 190.5 \text{ lbf/in}^2$$

Table 2A Dry;
Table 2.3.1

$$F'_g = F_g C_D C_t = \left(2020 \, \frac{\text{lbf}}{\text{in}^2}\right)(1.0)(1.0)$$
$$= 2020 \text{ lbf/in}^2$$

$$\theta = 71.56°$$

Eq. 3.10-1

$$F'_\theta = \frac{F'_g F'_{c\perp}}{F'_g \sin^2 \theta + F'_{c\perp} \cos^2 \theta}$$
$$= \frac{\left(2020.0 \, \frac{\text{lbf}}{\text{in}^2}\right)\left(781.2 \, \frac{\text{lbf}}{\text{in}^2}\right)}{\left(2020.0 \, \frac{\text{lbf}}{\text{in}^2}\right)(\sin^2 71.56°)}$$
$$ + \left(781.2 \, \frac{\text{lbf}}{\text{in}^2}\right)(\cos^2 71.56°)$$
$$= 832.3 \text{ lbf/in}^2$$

$$F'_\theta = 832.3 \, \frac{\text{lbf}}{\text{in}^2} > f_\theta = 190.5 \text{ lbf/in}^2$$
$$832.3 \text{ lbf/in}^2 > 190.5 \text{ lbf/in}^2$$
$$F'_\theta > f_\theta \quad \text{[OK]}$$

Beam Design: Glued Laminated Timber

Glulam members are fabricated from laminations that are $1^1/_2$ in thick with western species and $1^3/_8$ in thick with southern pine. NDS Supp. Tables 5A, 5B, and 5C contain all pertinent provisions for these specifications. For strength, grades of glulam members consist of two combinations: bending and axial.

1. Bending Combinations

[NDS 2.3.1, 3.2–3.5; and NDS Supp. Table 5A]

Members are stressed principally in bending but can be loaded axially, if necessary. However, since the bending combinations have higher-quality laminating wood at the outer fibers, they are more efficient as beams.

A combination symbol and the species of the laminating wood define the bending designation. For example, a combination symbol 24F-V3 indicates a combination with a tabulated bending stress of 2400 lbf/in² for a normal duration of loading (10 yr) and dry service conditions (16% or less moisture content). The letter V indicates visually graded lumber laminations. In addition, the species of wood is listed DF/DF, DF/HF, and so on, for outer lams/inner lams, with the outer lams a higher-quality lam wood and the inner lams a lower-quality lam wood.

Actual bending stresses are determined by
$$f_b = \frac{Mc}{I} = \frac{M}{S}$$

M = bending moment that uses the clear distance between the support faces plus one-half the required bearing length at each support for the span length

A. Volume Factor, C_V

[NDS 5.3.2; NDS Table 5.3.2]

When glued laminated timber is loaded perpendicular to the wide face of the laminations, tabulated bending design values for loading perpendicular to the wide faces of the laminations, F_{bxx}, shall be multiplied by the volume factor, C_V.

The tabulated design values for bending about the x-axis, F_{bxx} are based on a simple-span member $5^1/_8$ in wide, 12 in deep, and 21 ft in length that is loaded with a uniform load on the $5^1/_8$ in face. When a different member size is used or a different loading condition exists, the volume factor, C_V, is to be multiplied by the tabulated value.

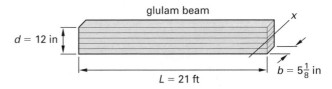

Figure 5.1 Base Dimensions for Bending Stress in Glulam

For western species glulam, the volume factor is given by

$$C_V = K_L \left(\frac{21}{L}\right)^{1/10} \left(\frac{12}{d}\right)^{1/10} \left(\frac{5.125}{b}\right)^{1/10} \leq 1.0$$

For southern pine glulam, the volume factor is given by

$$C_V = K_L \left(\frac{21}{L}\right)^{1/20} \left(\frac{12}{d}\right)^{1/20} \left(\frac{5.125}{b}\right)^{1/20} \leq 1.0$$

K_L (See Table 5.1 or NDS Table 5.3.2.)

L = length of beam between points of zero moment (ft)

d = depth of the beam (in)

b = width of the beam (in). For laminations that consist of more than one piece, b is the width of the widest piece in layup.

B. Beam Stability Factor, C_L

[NDS 3.3.3]

NDS Eq. 3.3-6 used previously for sawn lumber beams is also applicable to glulam beams.

Table 5.1 Loading Condition Coefficient, K_L, for Structural Glued Laminated Bending Members (NDS Table 5.3.2)

loading condition	K_L
single-span beam	
uniformly distributed load	1.0
concentrated load at midspan	1.09
two equal concentrated loads at one-third points of span	0.96
continuous beam or cantilever beam	
all loading conditions	1.0

Reproduced from *National Design Specification for Wood Construction*, 1997 Edition, courtesy of American Forest & Paper Association, Washington, D.C.

C. Allowable Bending Stress for Strong Axis (bending about x-axis)

A glulam beam bending combination stressed about the x-axis is usually with the tension laminations stressed in tension. The notation F_{bx} refers to this loading situation.

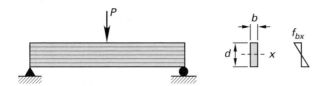

Figure 5.2 Glulam Beam Bending About the Strong Axis

The allowable bending stress is taken as the smaller of the following values.

With lateral stability,

$$F'_{bx} = F'_{bx,t/t} = F_{bx,t/t}C_DC_MC_tC_L$$

With volume effect,

$$F'_{bx} = F'_{bx,t/t} = F_{bx,t/t}C_DC_MC_tC_V$$

$F'_{bx} = F'_{bx,t/t}$ = allowable bending stress about the x-axis with high-quality tension laminations stressed in tension

$F_{bx} = F_{bx,t/t}$ = tabulated bending stress about the x-axis tension zone stressed in tension (See NDS Supp. Table 5A.)

C_D (See NDS Table 2.3.2; NDS App. B.)

$C_M = 1.0$ for MC $< 16\%$

$C_t = 1.0$ for normal temperature conditions (See NDS Table 2.3.4.)

$C_L = 1.0$ for continuous lateral support of compression face of beam. For other conditions of lateral support, C_L is evaluated in accordance with NDS 3.3.3.

$C_V = $ NDS 5.3.2 and NDS Eq. 5.3-1

Glulam beams are sometimes loaded in bending about the x-axis with the compression laminations stressed in tension. The typical application of this case is in a beam with a relatively short cantilever. The allowable bending stress is taken as the smaller of the following values.

With lateral stability,

$$F'_{bx,c/t} = F_{bx,c/t}C_DC_MC_tC_L$$

With volume effect,

$$F'_{bx,c/t} = F_{bx,c/t}C_DC_MC_tC_V$$

$F'_{bx,c/t} = $ allowable bending stress about the x-axis with compression laminations stressed in tension

$F_{bx,c/t} = $ tabulated bending stress about the x-axis with compression zone stressed in tension (see NDS Supp. Table 5A)

D. Allowable Bending Stress for Weak Axis (bending about y-axis)

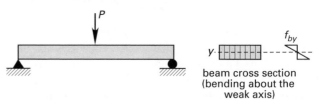

Figure 5.3 Glulam Beam Bending About the Weak Axis

The allowable bending stress is given by

$$F'_{by} = F'_{by}C_DC_MC_tC_{fu}$$

F_{by} (See NDS Supp. Table 5A.)

$C_{fu} = $ flat use factor (NDS 5.3.3). Obtain values from the Adj. Factors section of NDS Supp. Table 5A. Flat use factor may be taken conservatively as being equal to 1.0.

$C_V = $ not applicable

$C_L = 1.0$

2. Axial Combinations

[NDS Supp. Table 5B]

Members are stressed primarily in axial tension or compression. The combination symbols are numbered 1, 2, 3, and so on, followed by the wood species. This combination member will be more efficient as an axial member because the distribution of laminating wood grades is uniform across the member cross section.

The use of axial combination members as beams is allowable, although it is not as efficient as the bending combination beams. However, the same adjustment factors for the bending combination are applicable.

Example 5.1
Glulam Beam Analysis

A 32 ft glulam beam (5 in × 22 in, 24F-V1 southern pine bending combination) is laterally supported at 8 ft intervals (purlin points). Assume conditions of wet service and normal temperature (i.e., $C_t = 1.0$). The three loads ($P = 4000$ lbf each) consist of dead load and snow load (DL + SL), and the beam weight is 30 lbf/ft as illustrated. The total deflection may not exceed $L/240$.

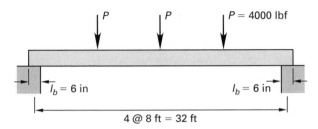

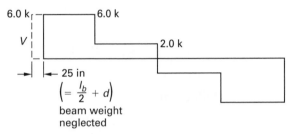

Review the beam.

Solution:

NDS Section

For 24F-V1 SP/SP glulam,

Supp.
Table 5A

	C_M
$F_{bx} = 2400$ lbf/in^2	0.80
$F_{vx} = 240$ lbf/in^2	0.875
$F_{c\perp x} = 740$ lbf/in^2	0.53
$E_x = 1,700,000$ lbf/in^2	0.833
$E_y = 1,500,000$ lbf/in^2	0.833

Table 2.3.2   For snow load, $C_D = 1.15$.

Supp.
Table 1C 5×22 glulam ($A = 110.0$ in^2, $S_x = 403.3$ in^3, and $I_x = 4436.7$ in^4)

A. Bending

Table 2.3.1   The smaller of the following values controls.
$$F'_{bx} = F'_{bx,t/t} = F_{bx,t/t}C_DC_MC_tC_L$$
$$= F^*_{bx,t/t}C_L$$
Alternatively,
$$F'_{bx} = F'_{bx,t/t} = F_{bx,t/t}C_DC_MC_tC_V$$
$$= F^*_{bx,t/t}C_V$$

3.3.3;
Table 3.3.3   Find the beam stability factor, C_L.

The unbraced length is
$$l_u = (8 \text{ ft})\left(12 \frac{\text{in}}{\text{ft}}\right) = 96 \text{ in}$$

The effective unbraced length is
$$l_e = 1.54l_u = (1.54)(96 \text{ in}) = 147.8 \text{ in}$$

Eq. 3.3-5 $$R_B = \sqrt{\frac{l_e d}{b^2}} = \sqrt{\frac{(147.8 \text{ in})(22 \text{ in})}{(5 \text{ in})^2}} = 11.4$$

3.3.3.8   For glulam beam, $K_{bE} = 0.609$.

Table 2.3.1 $$E' = E'_y = E_y C_M$$
$$= \left(1,500,000 \frac{\text{lbf}}{\text{in}^2}\right)(0.833)$$
$$= 1,249,500 \text{ lbf/in}^2$$

$$F_{bE} = \frac{K_{bE}E'_y}{R_B^2} = \frac{(0.609)\left(1,249,500 \frac{\text{lbf}}{\text{in}^2}\right)}{(11.4)^2}$$
$$= 5855 \text{ lbf/in}^2$$

$$F^*_{bx,t/t} = F_{bx,t/t}C_DC_MC_t$$
$$= \left(2400 \frac{\text{lbf}}{\text{in}^2}\right)(1.15)(0.80)(1.0)$$
$$= 2208.0 \text{ lbf/in}^2$$

Eq. 3.3-6   Let $F^*_b = F^*_{bx,t/t}$.

$$C_L = \frac{1 + \dfrac{F_{bE}}{F^*_b}}{1.9} - \sqrt{\left(\frac{1 + \dfrac{F_{bE}}{F^*_b}}{1.9}\right)^2 - \frac{\dfrac{F_{bE}}{F^*_b}}{0.95}}$$

$$= \frac{1 + \dfrac{5855 \frac{\text{lbf}}{\text{in}^2}}{2208 \frac{\text{lbf}}{\text{in}^2}}}{1.9}$$

$$- \sqrt{\left(\frac{1 + \dfrac{5855 \frac{\text{lbf}}{\text{in}^2}}{2208 \frac{\text{lbf}}{\text{in}^2}}}{1.9}\right)^2 - \frac{\dfrac{5855 \frac{\text{lbf}}{\text{in}^2}}{2208 \frac{\text{lbf}}{\text{in}^2}}}{0.95}}$$

$$= 0.972$$

Supp.
Table 5A
or Eq. 5.3-1   Find the volume factor of glulam, C_V.

For a nearly uniform load, $K_L = 1.0$.

For southern pine, $x = 20$.

$$C_V = K_L \left(\frac{21}{L}\right)^{1/x} \left(\frac{12}{d}\right)^{1/x} \left(\frac{5.125}{b}\right)^{1/x}$$

$$= (1.0) \left(\frac{21}{31 \text{ ft}}\right)^{1/20} \left(\frac{12}{22 \text{ in}}\right)^{1/20}$$

$$\times \left(\frac{5.125}{5 \text{ in}}\right)^{1/20}$$

$$= 0.951 \le 1.0$$

Since $C_L = 0.972 > C_V = 0.951$, use C_V.

$$F'_{bx} = F^*_{bx,t/t} C_V = \left(2208.0 \ \frac{\text{lbf}}{\text{in}^2}\right)(0.951)$$

$$= 2099.8 \text{ lbf/in}^2$$

$$M = M_{\text{beam}} + M_L \quad \text{[from shear diagram]}$$

$$= \frac{\left(30 \ \frac{\text{lbf}}{\text{ft}}\right)(32 \text{ ft})^2 \left(12 \ \frac{\text{in}}{\text{ft}}\right)}{8}$$

$$+ \left((6000 \text{ lbf})(8 \text{ ft}) + (2000 \text{ lbf})(8 \text{ ft})\right)$$

$$\times \left(12 \ \frac{\text{in}}{\text{ft}}\right)$$

$$= 814{,}080 \text{ in-lbf}$$

$$f_{bx} = \frac{M}{S_x} = \frac{814{,}080 \text{ in-lbf}}{403.3 \text{ in}^3}$$

$$= 2018.6 \text{ lbf/in}^2$$

$$f_{bx} = 2018.6 \ \frac{\text{lbf}}{\text{in}^2} < F'_{bx} = 2099.8 \text{ lbf/in}^2 \text{ [OK]}$$

3.4.3 and
3.4.3.1

B. Shear (accounting for the allowed disregard of loads at a distance from beam support equal to beam depth)

$$V_L = 6000 \text{ lbf} + \left(30 \ \frac{\text{lbf}}{\text{ft}}\right)$$

$$\times \left(\frac{32 \text{ ft} - \dfrac{(2)(22 \text{ in})}{12 \ \frac{\text{in}}{\text{ft}}}}{2}\right)$$

$$= 6425 \text{ lbf}$$

Eq. 3.4-2

$$f_v = (1.5)\left(\frac{V_L}{bd}\right) = (1.5)\left(\frac{6425 \text{ lbf}}{(5 \text{ in})(22 \text{ in})}\right)$$

$$= 87.6 \text{ lbf/in}^2$$

Table 2.3.1

$$F'_{vx} = F_{vx} C_D C_M C_t$$

$$= \left(240 \ \frac{\text{lbf}}{\text{in}^2}\right)(1.15)(0.875)(1.0)$$

$$= 241.5 \text{ lbf/in}^2$$

$$F'_{vx} = 241.5 \ \frac{\text{lbf}}{\text{in}^2} > f_v = 87.6 \text{ lbf/in}^2 \text{ [OK]}$$

2.3.10;
3.10.2

C. Beam bearing at support

The support reaction is

$$\left(30 \ \frac{\text{lbf}}{\text{ft}}\right)(16 \text{ ft}) + 4000 \text{ lbf} + 2000 \text{ lbf}$$

$$= 6480.0 \text{ lbf}$$

For $l_b = 6$ in or more, $C_b = 1.0$.

$$f_{c\perp} = \frac{P}{l_b b} = \frac{6480.0 \text{ lbf}}{(6 \text{ in})(5 \text{ in})}$$

$$= 216 \text{ lbf/in}^2$$

Table 2.3.1

$$F'_{c\perp} = F_{c\perp x} C_M C_t C_b$$

$$= \left(740 \ \frac{\text{lbf}}{\text{in}^2}\right)(0.53)(1.0)(1.0)$$

$$= 392.2 \ \frac{\text{lbf}}{\text{in}^2} > f_{c\perp} = 216 \text{ lbf/in}^2 \quad \text{[OK]}$$

D. Deflection

Table 2.3.1

$$E'_x = E_x C_M = \left(1{,}700{,}000 \ \frac{\text{lbf}}{\text{in}^2}\right)(0.833)$$

$$= 1{,}416{,}100 \text{ lbf/in}^2$$

$$\Delta = \frac{19PL^3}{384 E'_x I_x} + \frac{5wL^4}{384 E'_x I_x}$$

$$= \frac{\begin{array}{c}(19)(4000 \text{ lbf})\left((32 \text{ ft})\left(12 \ \frac{\text{in}}{\text{ft}}\right)\right)^3 \\[4pt] + (5)\left(30 \ \frac{\text{lbf}}{\text{ft}}\right)\left((32 \text{ ft})\left(12 \ \frac{\text{in}}{\text{ft}}\right)\right)^4\end{array}}{(384)\left(1{,}416{,}100 \ \frac{\text{lbf}}{\text{in}^2}\right)(4436.7 \text{ in}^4)}$$

$$= 3.14 \text{ in}$$

$$\frac{L}{240} = \frac{(32 \text{ ft})\left(12 \ \frac{\text{in}}{\text{ft}}\right)}{240}$$

$$= 1.6 \text{ in} < 3.14 \text{ in} \quad \text{[no good]}$$

Example 5.2
Glulam Beam Design

A glulam beam is a 24F-V3 SP/SP bending combination (NDS Table 5A). The design loads shown are caused by a combined DL + roof LL. The top face of the beam is laterally supported at the load points, and the bottom face in the negative moment area is laterally unsupported except at the reaction point. Assume wet service conditions at normal temperatures. The beam weight can be neglected for this example.

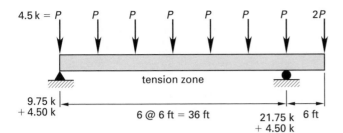

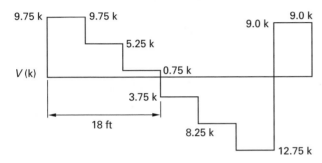

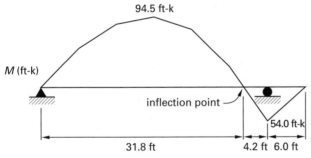

Go through a design procedure to determine the necessary glulam beam size considering bending and shear stresses only.

Solution:

NDS Section

For 24F-V3 SP/SP,

Supp.
Table 5A

	C_M
$F_{bx} = 2400 \text{ lbf/in}^2$	0.80
$F_{vx} = 240 \text{ lbf/in}^2$	0.875
$E_x = 1,800,000 \text{ lbf/in}^2$	0.833
$E_y = 1,600,000 \text{ lbf/in}^2$	0.833
$F_{bx} = 1200 \text{ lbf/in}^2$	0.80

(compression zone stressed in tension)

Table 2.3.2 For roof live load, $C_D = 1.25$ (for seven days construction live load).

A. Positive moment (tension zone stressed in tension).

Refer to shear and moment diagrams.

Table 2.3.1 The smaller of the following values controls.
$$F'_{bx} = F'_{bx,t/t} = F_{bx,t/t}C_DC_MC_tC_L$$
$$= F^*_{bx,t/t}C_L$$

Alternatively,
$$F'_{bx} = F'_{bx,t/t} = F_{bx,t/t}C_DC_MC_tC_V$$
$$= F^*_{bx,t/t}C_V$$

$$F^*_{bx,t/t} = F_{bx,t/t}C_DC_MC_t$$
$$= \left(2400 \, \frac{\text{lbf}}{\text{in}^2}\right)(1.25)(0.80)(1.0)$$
$$= 2400 \text{ lbf/in}^2$$

The initial beam size is
$$S = \frac{M}{F^*_{bx,t/t}}$$

$$= \frac{(94.5 \text{ ft-k})\left(1000 \, \frac{\text{lbf}}{\text{k}}\right)\left(12 \, \frac{\text{in}}{\text{ft}}\right)}{2400 \, \frac{\text{lbf}}{\text{in}^2}}$$

$$= 472.5 \text{ in}^3$$

Table 1C Try $5 \times 24^3/4$ ($A = 123.8 \text{ in}^2$; $S_x = 510.5 \text{ in}^3$).

Find the beam stability factor, C_L.

The unbraced length is
$$l_u = (6 \text{ ft})\left(12 \, \frac{\text{in}}{\text{ft}}\right) = 72 \text{ in}$$

Table 3.3.3 $\dfrac{l_u}{d} = \dfrac{72 \text{ in}}{24.75 \text{ in}} = 2.9 < 7.0$

The effective unbraced length is
$$l_e = 2.06l_u$$
$$= (2.06)(72 \text{ in}) = 148.32 \text{ in}$$

Eq. 3.3-5 and 3.3-6
$$R_B = \sqrt{\frac{l_e d}{b^2}} = \sqrt{\frac{(148.32 \text{ in})(24.75 \text{ in})}{(5 \text{ in})^2}}$$
$$= 12.12$$

3.3.3.8 For glulam beams, $K_{bE} = 0.609$.

Table 2.3.1
$$E'_y = E_yC_M = \left(1,600,000 \, \frac{\text{lbf}}{\text{in}^2}\right)(0.833)$$
$$= 1,332,800 \text{ lbf/in}^2$$

$$F_{bE} = \frac{K_{bE}E'_y}{R_B^2} = \frac{(0.609)\left(1,332,800 \, \frac{\text{lbf}}{\text{in}^2}\right)}{(12.12)^2}$$
$$= 5525.6 \text{ lbf/in}^2$$

$$\frac{F_{bE}}{F^*_{bx,t/t}} = \frac{5525.6 \ \frac{\text{lbf}}{\text{in}^2}}{2400.0 \ \frac{\text{lbf}}{\text{in}^2}}$$

$$= 2.30$$

$$C_L = \frac{1 + \dfrac{F_{bE}}{F^*_{bx,t/t}}}{1.9}$$

$$- \sqrt{\left(\dfrac{1 + \dfrac{F_{bE}}{F^*_{bx,t/t}}}{1.9}\right)^2 - \dfrac{\dfrac{F_{bE}}{F^*_{bx,t/t}}}{0.95}}$$

$$= \frac{1 + 2.30}{1.9} - \sqrt{\left(\frac{1+2.30}{1.9}\right)^2 - \frac{2.30}{0.95}}$$

$$= 0.96$$

Find the volume factor of glulam, C_V.

Eq. 5.3-1;
5.3.2;
Table 5.3.2

For a nearly uniform load, $K_L = 1.0$.

For southern pine, $x = 20$.

The distance between points of zero moment is $L = 31.8$ ft (see moment diagram).

$$C_V = K_L \left(\frac{21}{L}\right)^{1/x} \left(\frac{12}{d}\right)^{1/x} \left(\frac{5.125}{b}\right)^{1/x}$$

$$= (1.0) \left(\frac{21}{31.8 \text{ ft}}\right)^{1/20} \left(\frac{12}{24.75 \text{ in}}\right)^{1/20}$$

$$\times \left(\frac{5.125}{5 \text{ in}}\right)^{1/20}$$

$$= 0.946$$

$$0.946 < 0.96$$
$$C_V < C_L$$

Therefore, $C_V = 0.946$ controls.

$$F'_{bx} = F^*_{bx,t/t} C_V = \left(2400 \ \frac{\text{lbf}}{\text{in}^2}\right)(0.946)$$

$$= 2270.4 \ \text{lbf/in}^2$$

$$f_{bx} = \frac{M}{S_x} = \frac{(94.5 \text{ ft-k})\left(1000 \ \frac{\text{lbf}}{\text{k}}\right)\left(12 \ \frac{\text{in}}{\text{ft}}\right)}{510.5 \text{ in}^3}$$

$$= 2221.4 \ \frac{\text{lbf}}{\text{in}^2} < F'_{bx} \quad [\text{OK}]$$

B. Shear

$$f_v = (1.5)\left(\frac{V}{A}\right) = (1.5)\left(\frac{12,750 \text{ lbf}}{123.8 \text{ in}^2}\right)$$

$$= 154.5 \ \text{lbf/in}^2$$

Table 2.3.1

$$F'_{vx} = F_{vx} C_D C_M C_t$$

$$= \left(240 \ \frac{\text{lbf}}{\text{in}^2}\right)(1.25)(0.875)(1.0)$$

$$= 262.6 \ \text{lbf/in}^2$$

$$262.6 \ \text{lbf/in}^2 > 154.5 \ \text{lbf/in}^2$$
$$F'_{vx} > f_v \quad [\text{OK}]$$

C. Negative moment (compression zone stressed in tension)

The smaller of the following values controls.
$$F'_{bx} = F'_{bx,c/t} = F^*_{bx,c/t} C_L$$
Alternatively,
$$F'_{bx} = F'_{bx,c/t} = F^*_{bx,c/t} C_V$$

$$F^*_{bx,c/t} = F_{bx,c/t} C_D C_M C_t$$

$$= \left(1200 \ \frac{\text{lbf}}{\text{in}^2}\right)(1.25)(0.80)(1.0)$$

$$= 1200 \ \text{lbf/in}^2$$

For beam size 5 in $\times$ 24³⁄₄ in, $A = 123.8 \text{ in}^2$; $S_x = 510.5 \text{ in}^3$.

3.3.3

Find the beam stability factor, C_L.

The unbraced length is
$$l_u = (6 \text{ ft})\left(12 \ \frac{\text{in}}{\text{ft}}\right) = 72 \text{ in}$$

(Note: There are 4.2 ft between the support and the inflection point. This is less than 6 ft. Therefore, $l_u = 4.2$ ft is not critical.)

Table 3.3.3

$$\frac{l_u}{d} = \frac{72 \text{ in}}{24.75 \text{ in}} = 2.9 < 7.0$$

The effective unbraced length is
$$l_e = 1.87 l_u$$
$$= (1.87)(72 \text{ in}) = 134.64 \text{ in}$$

$$R_B = \sqrt{\frac{l_e d}{b^2}} = \sqrt{\frac{(134.64 \text{ in})(24.75 \text{ in})}{(5 \text{ in})^2}}$$

$$= 11.54$$

3.3.3.8

For glulam beams, $K_{bE} = 0.609$.

$$F_{bE} = \frac{K_{bE} E'_y}{R_B^2}$$

$$= \frac{(0.609)\left(1,332,800 \ \frac{\text{lbf}}{\text{in}^2}\right)}{(11.54)^2}$$

$$= 6094.96 \ \text{lbf/in}^2$$

$$\frac{F_{bE}}{F^*_{bx,c/t}} = \frac{6094.96 \ \frac{\text{lbf}}{\text{in}^2}}{1200 \ \frac{\text{lbf}}{\text{in}^2}}$$

$$= 5.08$$

$$C_L = \frac{1 + \dfrac{F_{bE}}{F^*_{bx,c/t}}}{1.9}$$

$$- \sqrt{\left(\frac{1 + \dfrac{F_{bE}}{F^*_{bx,c/t}}}{1.9}\right)^2 - \frac{\dfrac{F_{bE}}{F^*_{bx,c/t}}}{0.95}}$$

$$= \frac{1 + 5.08}{1.9} - \sqrt{\left(\frac{1 + 5.08}{1.9}\right)^2 - \frac{5.08}{0.95}}$$

$$= 0.99$$

Find the volume factor of glulam, C_V.

5.3.2;
Table 5.3.2;
Eq. 5.3-1

For cantilever, $K_L = 1.0$.

For southern pine, $x = 20$.

The distance between points of zero moment is $L = 4.2 \ \text{ft} + 6 \ \text{ft} = 10.2 \ \text{ft}$ (see the moment diagram).

$$\frac{1}{x} = \frac{1}{20}$$

$$C_V = K_L \left(\frac{21}{L}\right)^{1/x} \left(\frac{12}{d}\right)^{1/x} \left(\frac{5.125}{b}\right)^{1/x}$$

$$= (1.0) \left(\frac{21}{10.2 \ \text{ft}}\right)^{1/20} \left(\frac{12}{24.75 \ \text{in}}\right)^{1/20}$$

$$\times \left(\frac{5.125}{5 \ \text{in}}\right)^{1/20}$$

$$= 1.0$$

$$0.99 < 1.0$$

$$C_V < C_L$$

Therefore, $C_L = 0.99$ controls.

$$F'_{bx} = F'_{bx,c/t} = F^*_{bx,c/t} C_L$$

$$= \left(1200 \ \frac{\text{lbf}}{\text{in}^2}\right)(0.99)$$

$$= 1188 \ \text{lbf/in}^2$$

$$f_{bx} = \frac{M}{S_x}$$

$$= \frac{(54.0 \ \text{ft-k}) \left(1000 \ \frac{\text{lbf}}{\text{k}}\right) \left(12 \ \frac{\text{in}}{\text{ft}}\right)}{510.5 \ \text{in}^3}$$

$$= 1269.3 \ \frac{\text{lbf}}{\text{in}^2} > F'_{bx} = 1188 \ \text{lbf/in}^2$$

[no good]

$$\frac{f_{bx}}{F'_{bx}} = \frac{1269.3 \ \frac{\text{lbf}}{\text{in}^2}}{1188 \ \frac{\text{lbf}}{\text{in}^2}}$$

$$= 1.07 > 1.0 \quad [\text{no good}]$$

Strictly speaking, the beam analyzed is not satisfactory, and the design procedure must be repeated using a larger beam or higher-strength wood. Engineering judgment must be used to determine if 1.07 is close enough to 1.0 to allow use of this beam.

Example 5.3
Cantilevered Glulam Beam: Two Equal Spans

Roof girders consist of $5^{1}/_{8}$ in by $28^{1}/_{2}$ in glulam timbers. The combination symbol is 22F-V8 DF/DF (NDS Table 5A). The combined roof dead and snow load is 600 lbf/ft. The maximum allowable deflection is $L/180$.

Review the girders for bending and shear, and determine deflections at B (midpoint of span AB) and D (hinge point).

Solution:

NDS Section

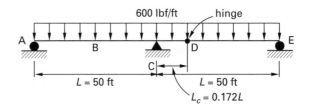

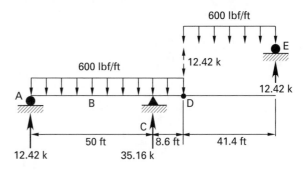

The shear is as follows.

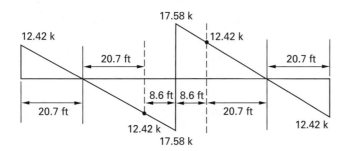

The moment diagram is as follows.

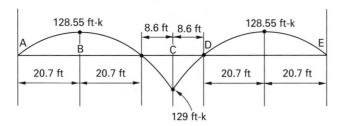

For $5^{1}/_{8}$ in $\times$ $28^{1}/_{2}$ in 22F-V8 DF/DF glulam,

NDS Supp.
Table 5A
$$C_M = C_t = 1.0$$
$$F_{bx} = 2200 \text{ lbf/in}^2$$
$$F_{vx} = 190 \text{ lbf/in}^2$$
$$E_x = 1.7 \times 10^6 \text{ lbf/in}^2$$
$$E_y = 1.6 \times 10^6 \text{ lbf/in}^2$$

This is a balanced glulam section. Thus, $F_{bx} = F_{bx,t/t} = F_{bx,c/t}$.

Table 2.3.2   For snow load, $C_D = 1.15$.

Suppl.
Table 1C
$$A = 146.1 \text{ in}^2$$
$$I_x = 9887.0 \text{ in}^4$$
$$S_x = 693.8 \text{ in}^3$$

Find the allowable bending stresses, F'_{bx}.

For member ABC (in the positive moment region), $C_L = 1.0$ since the roof sheathings and purlins provide lateral stability at the girder topside.

NDS Eq.
5.3-1 &
Suppl.
Table 5A
Adj. Factors
The volume factor is
$$C_V = K_L \left(\frac{21}{L}\right)^{1/10} \left(\frac{12}{d}\right)^{1/10}$$
$$\times \left(\frac{5.125}{b}\right)^{1/10} \leq 1.0$$
$$= (1.0) \left(\frac{21}{20.7 \text{ ft} + 20.7 \text{ ft}}\right)^{1/10}$$
$$\times \left(\frac{12}{28.5 \text{ in}}\right)^{1/10} \left(\frac{5.125}{5.125 \text{ in}}\right)^{1/10}$$
$$= 0.857$$

The allowable bending stress, F'_{bx}, for member ABC is

$$F'_{bx} = F_{bx}(C_D C_M C_t C_V)$$
$$= \left(2200 \frac{\text{lbf}}{\text{in}^2}\right)((1.15)(1.0)(1.0)(0.857))$$
$$= 2168.2 \text{ lbf/in}^2$$

For member BCD (negative moment region), assume the lateral support is provided by bracing the underside at support C.

Find the beam stability factor, C_L.

For the unbraced length,
$$l_u = (8.6 \text{ ft}) \left(12 \frac{\text{in}}{\text{ft}}\right) = 103.2 \text{ in}$$
$$\frac{l_u}{d} = \frac{103.2 \text{ in}}{28.5 \text{ in}} = 3.62 < 7.0$$

NDS
Table 3.3.3
The effective span length of the bending member is

$$l_e = 2.06 l_u = (2.06)(103.2 \text{ in}) = 212.6 \text{ in}$$

3.3.3.8
$$R_B = \sqrt{\frac{l_e d}{b^2}} = \sqrt{\frac{(212.6 \text{ in})(28.5 \text{ in})}{(5.125 \text{ in})^2}}$$

3.3.3.7
$$= 15.2 \leq 50 \quad [\text{OK}]$$
$$E'_y = E_y(C_M C_t)$$
$$= \left(1.6 \times 10^6 \frac{\text{lbf}}{\text{in}^2}\right)((1.0)(1.0))$$
$$= 1.6 \times 10^6 \text{ lbf/in}^2$$

3.3.3.8
$$F_{bE} = \frac{K_{bE} E'_y}{R_B{}^2}$$
For glulam, $K_{bE} = 0.610$.

$$F_{bE} = \frac{(0.610)\left(1.6 \times 10^6 \frac{\text{lbf}}{\text{in}^2}\right)}{(15.2)^2}$$
$$= 4224.4 \text{ lbf/in}^2$$
$$F^*_{bx} = F_{bx}(C_D C_M C_t)$$
$$= \left(2200 \frac{\text{lbf}}{\text{in}^2}\right)((1.15)(1.0)(1.0))$$
$$= 2530 \text{ lbf/in}^2$$

$$\frac{F_{bE}}{F^*_{bx}} = \frac{4224.4 \frac{\text{lbf}}{\text{in}^2}}{2530 \frac{\text{lbf}}{\text{in}^2}} = 1.67$$

$$C_L = \frac{1 + \dfrac{F_{bE}}{F_{bx}^*}}{1.9}$$

$$- \sqrt{\left(\frac{1 + \dfrac{F_{bE}}{F_{bx}^*}}{1.9}\right)^2 - \frac{\dfrac{F_{bE}}{F_{bx}^*}}{0.95}}$$

$$= \frac{1 + 1.67}{1.9} - \sqrt{\left(\frac{1 + 1.67}{1.9}\right)^2 - \frac{1.67}{0.95}}$$

$$= 0.940$$

Find the volume factor.

NDS
Eq. 5.3-1

$$C_V = K_L \left(\frac{21}{L}\right)^{1/x} \left(\frac{12}{d}\right)^{1/x} \left(\frac{5.125}{b}\right)^{1/x}$$

$$= (1.0) \left(\frac{21}{8.6 \text{ ft} + 8.6 \text{ ft}}\right)^{1/10}$$

$$\times \left(\frac{12}{28.5 \text{ in}}\right)^{1/10} \left(\frac{5.125}{5.125 \text{ in}}\right)^{1/10}$$

$$= 0.936 < 0.940 (= C_L)$$

Thus, $C_V = 0.936$ controls.

The allowable bending stress F_{bx}' for member BCD is

$$F_{bx}' = F_{bx}^* C_V = \left(2530 \ \frac{\text{lbf}}{\text{in}^2}\right)(0.936)$$

$$= 2368.1 \text{ lbf/in}^2$$

For member DE (in the positive moment region), the allowable bending stress is the same as for member AC.

$$F_{bx}' = 2168.2 \text{ lbf/in}^2$$

The allowable shear stress, F_v', is found as follows.

For all members,

$$F_v' = F_v(C_D C_M C_t)$$

$$= \left(190 \ \frac{\text{lbf}}{\text{in}^2}\right)((1.15)(1.0)(1.0))$$

$$= 218.5 \text{ lbf/in}^2$$

Compare the allowable stresses with the calculated stresses.

For members AC and DE,

$$f_b = \frac{M}{S_x}$$

$$= \frac{(128.55 \text{ ft-k})\left(1000 \ \dfrac{\text{lbf}}{\text{k}}\right)\left(12 \ \dfrac{\text{in}}{\text{ft}}\right)}{693.8 \text{ in}^3}$$

$$= 2223.4 \ \frac{\text{lbf}}{\text{in}^2} > F_{bx}' = 2168.2 \text{ lbf/in}^2$$

$$\text{[no good]}$$

$$f_v = \frac{3V}{2A} = \frac{(3)(12.42 \text{ k})\left(1000 \ \dfrac{\text{lbf}}{\text{k}}\right)}{(2)(146.1 \text{ in}^2)}$$

$$= 127.5 \ \frac{\text{lbf}}{\text{in}^2} < F_v' = 218.5 \text{ lbf/in}^2 \quad \text{[OK]}$$

For member BD,

$$f_b = \frac{M}{S_x} = \frac{(129 \text{ ft-k})\left(1000 \ \dfrac{\text{lbf}}{\text{k}}\right)\left(12 \ \dfrac{\text{in}}{\text{ft}}\right)}{693.8 \text{ in}^3}$$

$$= 2231.2 \ \frac{\text{lbf}}{\text{in}^2} < F_{bx}' = 2368.1 \text{ lbf/in}^2 \quad \text{[OK]}$$

$$f_v = \frac{3V}{2A} = \frac{(3)(17.58 \text{ k})\left(1000 \ \dfrac{\text{lbf}}{\text{k}}\right)}{(2)(146.1 \text{ in}^2)}$$

$$= 180.5 \ \frac{\text{lbf}}{\text{in}^2} < F_v' = 218.5 \text{ lbf/in}^2 \quad \text{[OK]}$$

Determine deflections at B (midpoint of span AC) and at hinge point D for camber purposes.

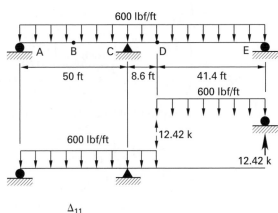

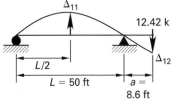

$$\Delta_{11} = \left(\frac{Pax}{6E_x'IL}\right)(L^2 - x^2)$$

$$= \frac{(12.42 \text{ k})(8.6 \text{ ft})(25 \text{ ft})}{(6)\left(1.7 \times 10^3 \frac{\text{k}}{\text{in}^2}\right)(9887 \text{ in}^4)(50 \text{ ft})}$$

$$\times \left((50 \text{ ft})^2 - (25 \text{ ft})^2\right)\left(\frac{(12 \text{ in})^3}{(1 \text{ ft})^3}\right)$$

$$= 1.72 \quad [\text{upward}]$$

$$= -1.72 \text{ in}$$

$$\Delta_{12} = \left(\frac{Pa^2}{3E_x'I_x}\right)(L + a)$$

$$= \left(\frac{(12.42 \text{ k})(8.6 \text{ ft})^2}{(3)\left(1.7 \times 10^3 \frac{\text{k}}{\text{in}^2}\right)(9887 \text{ in}^4)}\right)$$

$$\times (50 \text{ ft} + 8.6 \text{ ft})\left(\frac{(12 \text{ in})^3}{(1 \text{ ft})^3}\right)$$

$$= 1.84 \text{ in} \quad [\text{downward}]$$

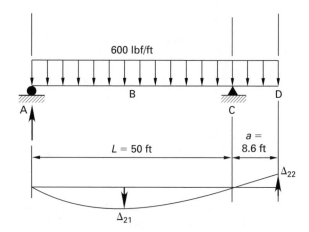

$$\Delta_{22} = \left(\frac{wa}{24E_x'I}\right)(4a^2L - L^3 + 3a^3)$$

$$= \left(\frac{\left(600 \frac{\text{lbf}}{\text{ft}}\right)\left(\frac{1 \text{ k}}{1000 \text{ lbf}}\right)(8.6 \text{ ft})}{(24)\left(1.7 \times 10^3 \frac{\text{k}}{\text{in}^2}\right)(9887 \text{ in}^4)}\right)$$

$$\times \left((4)(8.6 \text{ ft})^2(50 \text{ ft}) - (50 \text{ ft})^3 + (3)\right.$$

$$\left.\times (8.6 \text{ ft})^3\right)\left(\frac{(12 \text{ in})^3}{(1 \text{ ft})^3}\right)$$

$$= 2.39 \text{ in} \quad [\text{upward}]$$

$$= -2.39 \text{ in}$$

$$\Delta_{21} = \left(\frac{wx}{24E_x'IL}\right)$$

$$\times \left(L^4 - 2L^2x^2 + Lx^3 - 2a^2L^2 + 2a^2x^2\right)$$

$$= \left(\frac{\left(0.6 \frac{\text{k}}{\text{ft}}\right)(25 \text{ ft})}{(24)\left(1.7 \times 10^3 \frac{\text{k}}{\text{in}^2}\right)(9887 \text{ in}^4)(50 \text{ ft})}\right)$$

$$\times \left((50 \text{ ft})^4 - (2)(50 \text{ ft})^2(25 \text{ ft})^2\right.$$

$$+ (50 \text{ ft})(25 \text{ ft})^3 - (2)(8.6 \text{ ft})^2(50 \text{ ft})^2$$

$$\left.+ (2)(8.6 \text{ ft})^2(25 \text{ ft})^2\right)\left(\frac{12 \text{ in}}{1 \text{ ft}}\right)^3$$

$$= 4.66 \text{ in} \quad [\text{downward}]$$

The deflection at B (midpoint of AC) is
Δ_{11} upward $+\Delta_{21}$ downward

$$= -1.72 \text{ in} + 4.66 \text{ in}$$

$$= 2.94 \text{ in} \quad [\text{downward}]$$

The deflection at D (hinge point) is
Δ_{12} downward $+\Delta_{22}$ upward

$$= 1.84 \text{ in} - 2.39 \text{ in}$$

$$= -0.55 \text{ in} \quad [\text{upward}]$$

Axial Members, and Combined Bending and Axial Loading

1. Tension Members

[NDS 2.3.1 and 3.8; NDS Supp. Tables 4A, 4B, 4D, 5A, 5B, and 5C]

The actual tensile stress parallel to grain is based on the net section area and shall not exceed the allowable tensile design value parallel to grain.

$$f_t = \frac{P}{A_n} \le F'_t$$

P = tensile force

$F'_t = F_t C_D C_M C_t C_F C_i$ [see NDS Table 2.3.1]

F_t (See NDS Supp. Tables 4A–4D and 5A–5C.)

C_D (See NDS Table 2.3.2 and App. B.)

C_M = wet service factor < 1.0 when
MC $> 19\%$ for sawn lumber
MC $> 16\%$ for glulam
(See NDS Supp. Tables 4 and 5 Adj. Factors.)
= 1.0 for dry service conditions

C_F = size factor (NDS Table 2.3.1 and 4.3.2). For visually graded dimension lumber (i.e., 2 in to 4 in thick), see NDS Supp. Tables 4A, 4B, 4D, and 4E, Adj. Factors.

C_t = 1.0 for $T \le 100°F$
C_i = incising factor for sawn lumber (See NDS 2.3.11.)

Example 6.1
Tension Member: Sawn Lumber

Assume no. 2 hem-fir, 2×8 tension member with a row of $^3/_4$ in diameter bolts with $^1/_8$ in oversized holes; MC $> 19\%$; $P = DL + SL = 6000$ lbf.

Find f_t and F'_t.

Solution:
NDS Section

$$f_t = \frac{P}{A_n}$$
$$= \frac{6000 \text{ lbf}}{(1.5 \text{ in})(7.25 \text{ in} - (0.75 \text{ in} + 0.125 \text{ in}))}$$
$$= 627.5 \text{ lbf/in}^2$$

Supp.	$F_t = 525 \text{ lbf/in}^2$
Table 4A and Adj. Factors	$C_F = 1.2$
	$C_M = 1.0$
	$C_t = 1.0$
Table 2.3.2	$C_D = 1.15$

Table 2.3.1 $F'_t = F_t C_D C_M C_t C_F$

$$= \left(525 \frac{\text{lbf}}{\text{in}^2}\right)(1.15)(1.0)(1.0)(1.2)$$
$$= 724.5 \text{ lbf/in}^2$$

Example 6.2
Tension Member: Glulam

Glulam axial combination 5DF $2^1/_2 \times 6$ (4 lams; $A = 15.0 \text{ in}^2$), with a row of $^3/_4$ in diameter bolts with $^1/_8$ in over-sized holes; MC $> 19\%$; $P = DL + SL = 6000$ lbf.

Find f_t and F'_t.

Solution:
NDS Section

Supp.
Table 1C $2^1/_2$ in $\times$ 6 in are net finished dimensions.

$$f_t = \frac{P}{A_n}$$
$$= \frac{6000 \text{ lbf}}{(2.5 \text{ in})(6.0 \text{ in} - (0.75 \text{ in} + 0.125 \text{ in}))}$$
$$= 468 \text{ lbf/in}^2$$

Supp.   For 6DF axial combination, $F_t = 1600 \text{ lbf/in}^2$.

Table 5B and Adj. Factors $C_M = 0.80$

Table 2.3.2 $C_D = 1.15$

Table 2.3.1 $F'_t = F_t C_D C_M C_t$

$$= \left(1600 \frac{\text{lbf}}{\text{in}^2}\right)(1.15)(0.80)(1.0)$$
$$= 1472 \text{ lbf/in}^2$$

2. Combined Tension and Bending

[NDS 3.9.1, Eqs. 3.9-1 and 3.9-2]

Members subjected to a combination of axial tension and bending will satisfy NDS Eqs. 3.9-1 and 3.9-2.

$$\frac{f_t}{F_t'} + \frac{f_b}{F_b^*} \leq 1.0 \qquad \text{[NDS Eq. 3.9-1]}$$

f_b = actual bending tensile stress

F_b^* = allowable bending tensile stress without the stability factor, C_L

$\quad = F_b C_D C_M C_t C_F$ for sawn lumber

$\quad = F_b C_D C_M C_t C_V$ for glulam

To check the buckling on the compression face,

$$\frac{f_b - f_t}{F_b^{**}} \leq 1.0 \quad \text{[if } f_b > f_t\text{]} \qquad \text{[NDS Eq. 3.9-2]}$$

f_b = actual bending compressive stress

F_b^{**} = allowable bending compressive stress without the volume factor, C_V

$\quad = F_b C_D C_M C_t C_L C_F$ for sawn lumber

$\quad = F_b C_D C_M C_t C_L$ for glulam

Example 6.3
Combined Tension and Bending: Sawn Lumber

Given a no. 2 hem-fir 2×8 member with a row of $^3/_4$ in diameter bolts; MC > 19%; $P = \text{DL} + \text{SL} = 6000$ lbf; $M = 10{,}000$ in-lbf, review the member.

Solution:

NDS Section

$$f_t = \frac{P}{A_n}$$

$$\quad = 627.5 \text{ lbf/in}^2 \quad \text{[from Ex. 6.1]}$$

$$F_t' = 724.5 \text{ lbf/in}^2 \quad \text{[from Ex. 6.1]}$$

$$S = \frac{(1.5 \text{ in})(7.25 \text{ in})^2}{6} = 13.14 \text{ in}^3$$

$$f_b = \frac{M}{S} = \frac{10{,}000 \text{ in-lbf}}{13.14 \text{ in}^3}$$

$$\quad = 761.0 \text{ lbf/in}^2 \quad \begin{bmatrix} \text{either tensile} \\ \text{or compressive} \end{bmatrix}$$

Table 2.3.1 $\quad F_b^* = F_b C_D C_M C_t C_F$

Supp.
Table 4A
and Adj. $\quad F_b = 850 \text{ lbf/in}^2$

$\quad C_F = 1.2$

Factors $\quad$ Since $F_b C_F = 1020 \text{ lbf/in}^2 < 1150 \text{ lbf/in}^2$,

$\quad C_M = 1.0$

Table 2.3.2 $\quad C_D = 1.15$

$C_t = 1.0$

$$F_b^* = \left(850 \frac{\text{lbf}}{\text{in}^2} \right)(1.15)(1.0)(1.0)(1.2)$$

$$\quad = 1173.0 \text{ lbf/in}^2$$

Eq. 3.9-1
$$\frac{f_t}{F_t'} + \frac{f_b}{F_b^*} = \frac{627.5 \dfrac{\text{lbf}}{\text{in}^2}}{724.5 \dfrac{\text{lbf}}{\text{in}^2}} + \frac{761.0 \dfrac{\text{lbf}}{\text{in}^2}}{1173.0 \dfrac{\text{lbf}}{\text{in}^2}}$$

$$\quad = 1.515 > 1.0 \quad \text{[no good]}$$

Assuming the beam stability factor, C_L, is 1.0,

$$F_b^{**} = F_b C_D C_M C_t C_L C_F = F_b^* C_L$$

$$\quad = \left(1173 \frac{\text{lbf}}{\text{in}^2} \right)(1.0)$$

$$\quad = 1173 \text{ lbf/in}^2$$

Eq. 3.9-2
$$\frac{f_b - f_t}{F_b^{**}} = \frac{761 \dfrac{\text{lbf}}{\text{in}^2} - 627.5 \dfrac{\text{lbf}}{\text{in}^2}}{1173.0 \dfrac{\text{lbf}}{\text{in}^2}}$$

$$\quad = 0.114 < 1.0 \quad \text{[OK]}$$

Therefore, 2×8 no. 2 hem-fir is no good.

Example 6.4
Combined Tension and Bending: Glulam

A 30 ft long glulam tension member, axial combination 3DF (NDS Supp. Table 5B) $5^1/_8 \times 10^1/_2$, is subjected to a combination of 25,000 lbf tension and a bending moment of 80,000 in-lbf about the x-axis (caused by a uniformly distributed load). Assume MC > 16%, normal temperature ($C_t = 1.0$), and normal load duration ($C_D = 1.0$). Lateral supports are provided at the member ends only.

Check the adequacy of the member.

Solution:

NDS Section

Supp.
Table 1C
(DF) $\quad$ For $5^1/_8$ in $\times$ $10^1/_2$ in western species with seven laminations ($10^1/_2$ in/$1^1/_2$ in),

$$A = 53.81 \text{ in}^2$$

$$S_x = 94.17 \text{ in}^3$$

Supp.
Table 5B and
Table 5B Adj. $\quad$ For 3DF axial combination,

$$E_y = 1{,}800{,}000 \text{ lbf/in}^2; \ C_M = 0.833$$

Factors $\quad F_t = 1450 \text{ lbf/in}^2; \ C_M = 0.80$

$$F_{bx} = 2300 \text{ lbf/in}^2; \ C_M = 0.80$$

A. Axial

$$f_t = \frac{P}{A_n} = \frac{25{,}000 \text{ lbf}}{53.81 \text{ in}^2}$$

$$= 464.6 \text{ lbf/in}^2 \quad \text{[tensile or compressive]}$$

Table 2.3.1
$$F_t' = F_t C_D C_M C_t$$

$$= \left(1450 \, \frac{\text{lbf}}{\text{in}^2}\right)(1.0)(0.80)(1.0)$$

$$= 1160.0 \text{ lbf/in}^2$$

B. Bending

$$f_{bx} = \frac{M}{S_x} = \frac{80{,}000 \text{ in-lbf}}{94.17 \text{ in}^3}$$

$$= 849.5 \text{ lbf/in}^2$$

C. Combined axial tension and bending

5.3.2
The volume factor is given by

$$C_V = K_L \left(\frac{21}{L}\right)^{1/x} \left(\frac{12}{d}\right)^{1/x} \left(\frac{5.125}{b}\right)^{1/x}$$
$$\leq 1.0$$

Table 5.3.2
For a uniformly distributed load, $K_L = 1.0$.

For all species other than southern pine, $x = 10$.

$$C_V = (1.0) \left(\frac{21}{30 \text{ ft}}\right)^{1/10} \left(\frac{12}{10.5 \text{ in}}\right)^{1/10}$$

$$\times \left(\frac{5.125}{5.125 \text{ in}}\right)^{1/10}$$

$$= 0.978 < 1.0$$

3.9.1
$$F_{bx}^* = F_{bx} C_D C_M C_t C_V$$

$$= \left(2300 \, \frac{\text{lbf}}{\text{in}^2}\right)(1.0)(0.8)(1.0)(0.978)$$

$$= 1799.5 \text{ lbf/in}^2$$

Eq. 3.9-1
$$\frac{f_t}{F_t'} + \frac{f_{bx}}{F_{bx}^*} = \frac{464.6 \, \frac{\text{lbf}}{\text{in}^2}}{1160.0 \, \frac{\text{lbf}}{\text{in}^2}} + \frac{849.5 \, \frac{\text{lbf}}{\text{in}^2}}{1799.5 \, \frac{\text{lbf}}{\text{in}^2}}$$

$$= 0.87 < 1.0 \quad \text{[OK]}$$

3.9.1
$$F_{bx}^{**} = F_b C_D C_M C_t C_L = F_{bx}^* C_L$$

3.3.3.8
Find the beam stability factor, C_L.

Eq. 3.3-6;
Table 3.3.3
For the unbraced length,

$$l_u = (30 \text{ ft}) \left(12 \, \frac{\text{in}}{\text{ft}}\right) = 360 \text{ in}$$

$$\frac{l_u}{d} = \frac{360 \text{ in}}{10.5 \text{ in}} = 34.3 > 7$$

For the effective unbraced length for uniformly distributed load,

$$l_e = 1.63 l_u + 3d$$

$$= (1.63)(360 \text{ in}) + (3)(10.5 \text{ in})$$

$$= 618.3 \text{ in}$$

Eq. 3.3-5
$$R_B = \sqrt{\frac{l_e d}{b^2}} = \sqrt{\frac{(618.3 \text{ in})(10.5 \text{ in})}{(5.125 \text{ in})^2}}$$

$$= 15.72$$

3.3.3
For glulam, $K_{bE} = 0.610$.

$$E_y' = E_y C_M C_t$$

$$= \left(1{,}800{,}000 \, \frac{\text{lbf}}{\text{in}^2}\right)(0.833)(1.0)$$

$$= 1{,}499{,}400 \text{ lbf/in}^2$$

$$F_{bE} = \frac{K_{bE} E'}{R_B^2}$$

$$= \frac{(0.61)\left(1{,}499{,}400 \, \frac{\text{lbf}}{\text{in}^2}\right)}{(15.72)^2}$$

$$= 3701.1 \text{ lbf/in}^2$$

3.3.3.8
$$F_{bx}^* = F_b C_D C_M C_t$$

$$= \left(2300 \, \frac{\text{lbf}}{\text{in}^2}\right)(1.0)(0.8)(1.0)$$

$$= 1840 \text{ lbf/in}^2$$

$$\frac{F_{bE}}{F_{bx}^*} = \frac{3701.1 \, \frac{\text{lbf}}{\text{in}^2}}{1840 \, \frac{\text{lbf}}{\text{in}^2}} = 2.01$$

$$\frac{1 + \frac{F_{bE}}{F_{bx}^*}}{1.9} = \frac{1 + 2.01}{1.9} = 1.58$$

Eq. 3.3-6
$$C_L = \frac{1 + \frac{F_{bE}}{F_{bx}^*}}{1.9} - \sqrt{\left(\frac{1 + \frac{F_{bE}}{F_{bx}^*}}{1.9}\right)^2 - \frac{\frac{F_{bE}}{F_{bx}^*}}{0.95}}$$

$$= 1.58 - \sqrt{(1.58)^2 - \frac{2.01}{0.95}}$$

$$= 0.96$$

3.9.1
$$F_{bx}^{**} = F_{bx}^* C_L$$

$$= \left(1840 \, \frac{\text{lbf}}{\text{in}^2}\right)(0.96)$$

$$= 1766.4 \text{ lbf/in}^2$$

Eq. 3.9-2   If the bending compressive stress, f_b, exceeds the axial tensile stress, f_t,

$$\frac{f_{bx} - f_t}{F_{bx}^{**}} = \frac{849.5 \ \frac{\text{lbf}}{\text{in}^2} - 464.6 \ \frac{\text{lbf}}{\text{in}^2}}{1766.4 \ \frac{\text{lbf}}{\text{in}^2}}$$

$$= 0.22 < 1.0 \quad [\text{OK}]$$

Therefore, a 30 ft long glulam tension member, 3DF, $5\frac{1}{8}$ in $\times$ $10\frac{1}{2}$ in, is adequate.

3. Compression Members (Columns)

[NDS 3.7]

A column is a compression member. Although only solid columns will be treated in this book, columns are classified as follows.

- *Solid Columns*

[NDS 3.6 and 3.7]

Columns are relatively heavy vertical members.

Struts are relatively smaller compression members, not necessarily in a vertical position, for example, trench shores.

Studs are light compression members in wood framing, for example, wall studs.

Truss compression members are those members of a truss that resist compressive forces.

- *Spaced Columns*

[NDS 15.2]

Spaced columns are formed from two or more individual members that are spaced apart but act as a composite member.

- *Built-Up Columns*

[NDS 3.6.3, 3.7, and 15.3]

Built-up columns are individual laminations that are joined mechanically with no spaces between laminations.

A *long column* is a column that will buckle before it reaches the crushing capacity of the wood (maximum compressive stress parallel to grain). A *short column*, however, is a column that will not buckle, and its strength is limited to the crushing capacity of the wood.

The check on the load capacity of a column is given by

$$f_c = \frac{P}{A} \le F_c'$$

$F_c' = F_c C_D C_M C_t C_F C_i C_P$
(See Table 3.2 or NDS Table 2.3.1.)

F_c (See NDS Supp. Tables 4 and 5.)

C_P = column stability factor (See NDS 3.7.1.)
= 1.0 for columns that are fully braced (laterally supported) for the entire column length

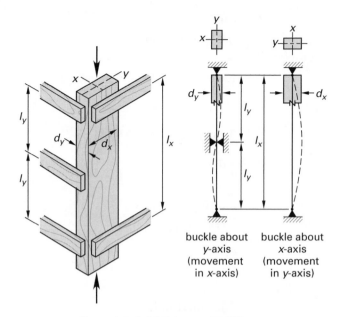

Figure 6.1 Solid Column with Different Unbraced Lengths for Both Axes

The column stability factor is given by

$$C_P = \frac{1 + \frac{F_{cE}}{F_c^*}}{2c} - \sqrt{\left(\frac{1 + \frac{F_{cE}}{F_c^*}}{2c}\right)^2 - \frac{\frac{F_{cE}}{F_c^*}}{c}}$$

[NDS Eq. 3.7-1]

F_{cE} = Euler critical buckling stress for columns
$$= \frac{K_{cE} E'}{\left(\frac{l_e}{d}\right)^2}$$

l = unbraced column length

l_e = effective unbraced column length $= K_e l$

K_e = buckling length coefficient (See NDS App. G.)

F_c^* = allowable crushing strength (limiting compressive strength in column at zero slenderness ratio)
= tabulated compressive stress parallel to grain multiplied by all adjustment factors except C_P
$= F_c C_D C_M C_t C_F C_i$

K_{cE} = Euler buckling coefficient
= 0.3 for visually graded lumber
= 0.418 for MSR lumber and glulam

E' = modulus of elasticity associated with the axis of column buckling. Recall that C_D does not apply to E. For sawn lumber, $E = E_x = E_y$. For glulam, E_x and E_y may be different.
$= E C_M C_t C_i C_T$

c = buckling and crushing interaction factor for columns

= 0.9 for glulam columns

= 0.8 for sawn lumber columns

C_T = buckling stiffness factor for 2×4 or smaller truss compression chords subjected to combined flexure and compression with plywood nailed to narrow face of member (See NDS 4.4.3.)

= 1.0 for all other members

C_F = size factor (NDS 4.3.2 and Table 2.3.1 footnote) for compression. Obtain values for visually graded dimension lumber (NDS Supp. Tables 4A, 4B, 4D, and 4E).

= not applicable to glulam (See NDS Table 2.3.1.)

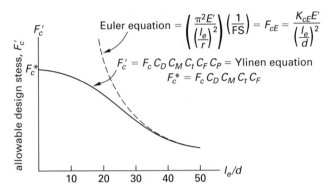

Figure 6.2 *Column Design Stress Versus* l_e/d

Example 6.5
Axially Loaded Column: Sawn Lumber

No. 3 SPF 2×4 bearing wall stud supports are shown in the following illustration. The total load is $P = \text{DL} + \text{SL} = 1600$ lbf. Assume dry service and normal temperature conditions ($C_M = 1.0$, $C_t = 1.0$).

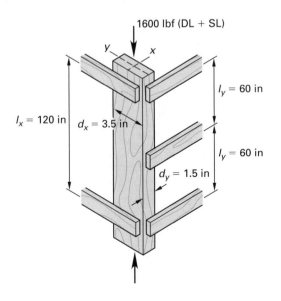

Check the stud capacity.

Solution:

NDS Section

$$f_c = \frac{P}{A} = \frac{1600 \text{ lbf}}{(1.5 \text{ in})(3.5 \text{ in})}$$
$$= 304.8 \text{ lbf/in}^2$$

Table 2.3.1 $\quad F_c' = F_c C_D C_M C_t C_F C_P = F_c^* C_P$

Table 2.3.2 $\quad$ For snow, $C_D = 1.15$.

Supp.
Table 4A; $\quad F_c = 650 \text{ lbf/in}^2$; $C_F = 1.15$; $C_M = 1.0$
Adj. Factors
$\quad C_t = 1.0$

$$F_c^* = F_c C_D C_M C_t C_F$$
$$= \left(650 \frac{\text{lbf}}{\text{in}^2}\right)(1.15)(1.0)(1.0)(1.15)$$
$$= 859.7 \text{ lbf/in}^2$$

$$E' = E C_M C_t$$
$$= \left(1{,}200{,}000 \frac{\text{lbf}}{\text{in}^2}\right)(1.0)(1.0)$$
$$= 1{,}200{,}000 \text{ lbf/in}^2$$

3.7; $\quad$ Find the column stability factor, C_P.
Eq. 3.7-1
$$\left(\frac{l_e}{d}\right)_x = \left(\frac{K_e l}{d}\right)_x = \frac{(1.0)(120 \text{ in})}{3.5 \text{ in}} = 34.3$$

$$\left(\frac{l_e}{d}\right)_y = \left(\frac{K_e l}{d}\right)_y = \frac{(1.0)(60 \text{ in})}{1.5 \text{ in}} = 40.0$$
$$= \left(\frac{l_e}{d}\right)_{\max} \quad \text{[controls]}$$

3.7.1 $\quad$ For visually graded lumber, $K_{cE} = 0.3$.

For sawn lumber, $c = 0.8$.

$$F_{cE} = \frac{K_{cE} E'}{\left(\frac{l_e}{d}\right)^2}$$
$$= \frac{(0.3)\left(1{,}200{,}000 \frac{\text{lbf}}{\text{in}^2}\right)}{(40.0)^2}$$
$$= 225.0 \text{ lbf/in}^2$$

$$\frac{F_{cE}}{F_c^*} = \frac{225.0 \frac{\text{lbf}}{\text{in}^2}}{859.7 \frac{\text{lbf}}{\text{in}^2}} = 0.262$$

$$\frac{1 + \frac{F_{cE}}{F_c^*}}{2c} = \frac{1 + 0.262}{(2)(0.8)} = 0.789$$

$$C_P = \frac{1 + \dfrac{F_{cE}}{F_c^*}}{2c} - \sqrt{\left(\frac{1 + \dfrac{F_{cE}}{F_c^*}}{2c}\right)^2 - \frac{\dfrac{F_{cE}}{F_c^*}}{c}}$$

$$= 0.789 - \sqrt{(0.789)^2 - \frac{0.262}{0.8}}$$

$$= 0.246$$

$$F_c' = F_c^* C_P$$

$$= \left(859.7 \ \frac{\text{lbf}}{\text{in}^2}\right)(0.246)$$

$$= 211.5 \ \text{lbf/in}^2$$
$$211.5 \ \text{lbf/in}^2 < 304.8 \ \text{lbf/in}^2$$
$$F_c' < f_c \quad [\text{no good}]$$

Example 6.6
Axially Loaded Column: Glulam

Assume an axial combination southern pine 47 glulam 5×11 column. Assume MC > 16% and normal temperature condition.

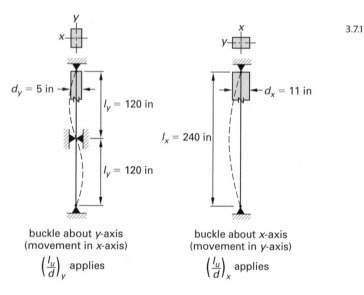

buckle about y-axis   buckle about x-axis
(movement in x-axis)   (movement in y-axis)

$\left(\dfrac{l_u}{d}\right)_y$ applies $\left(\dfrac{l_u}{d}\right)_x$ applies

Find the allowable compressive stress, F_c', for DL + SL loading.

Solution:

NDS Section

Supp.
Table 1C

For 5×11, eight lams,

$$A = 55.0 \ \text{in}^2; \ I_x = 554.6 \ \text{in}^4; \ I_y = 114.6 \ \text{in}^4$$

Table 2.3.1 $F_c' = F_c C_D C_M C_t C_P$

Supp.
Table 5B

$$F_c = 1900 \ \text{lbf/in}^2; \ C_M = 0.73$$

Supp.
Table 5B
Adj. Factors

Note that C_F is not applicable for glulam. (See NDS Table 2.3.1.)

$$E = 1,400,000 \ \text{lbf/in}^2; \ C_M = 0.833$$

$$F_c^* = F_c C_D C_M C_t$$

$$= \left(1900 \ \frac{\text{lbf}}{\text{in}^2}\right)(1.15)(0.73)(1.0)$$

$$= 1595.0 \ \text{lbf/in}^2$$

$$E' = E C_M C_t$$

$$= \left(1,400,000 \ \frac{\text{lbf}}{\text{in}^2}\right)(0.833)(1.0)$$

$$= 1,166,200 \ \text{lbf/in}^2$$

$l_e = K_e l_u$ and $K_e = 1.0$; $l_u = l_x$ or l_y

$$\left(\frac{l_e}{d}\right)_x = \frac{(1.0)(240 \ \text{in})}{11 \ \text{in}} = 21.8$$

$$\left(\frac{l_e}{d}\right)_y = \frac{(1.0)(120 \ \text{in})}{5 \ \text{in}} = 24.0$$

$$= \left(\frac{l_e}{d}\right)_{\text{max}} \quad [\text{controls}]$$

3.7.1   For glulam, $K_{cE} = 0.418$ and $c = 0.9$.

$$F_{cE} = \frac{K_{cE} E'}{\left(\dfrac{l_e}{d}\right)^2}$$

$$= \frac{(0.418)\left(1,166,200 \ \dfrac{\text{lbf}}{\text{in}^2}\right)}{(24)^2}$$

$$= 846.3 \ \text{lbf/in}^2$$

$$\frac{F_{cE}}{F_c^*} = \frac{846.3 \ \dfrac{\text{lbf}}{\text{in}^2}}{1595 \ \dfrac{\text{lbf}}{\text{in}^2}} = 0.53$$

$$\frac{1 + \dfrac{F_{cE}}{F_c^*}}{2c} = \frac{1 + 0.53}{(2)(0.9)} = 0.85$$

$$C_P = \frac{1 + \dfrac{F_{cE}}{F_c^*}}{2c} - \sqrt{\left(\frac{1 + \dfrac{F_{cE}}{F_c^*}}{2c}\right)^2 - \frac{\dfrac{F_{cE}}{F_c^*}}{c}}$$

$$= 0.85 - \sqrt{(0.85)^2 - \frac{0.53}{0.9}}$$

$$= 0.484$$

$$F_c' = F_c^* C_P$$

$$= \left(1595 \ \frac{\text{lbf}}{\text{in}^2}\right)(0.484)$$

$$= 772 \ \text{lbf/in}^2$$

4. Combined Compression and Bending

[NDS 3.9.2, NDS Eq. 3.9-3]

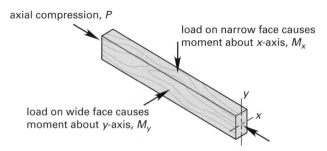

Figure 6.3 *Combined Compression and Bending About x- and y-Axes*

Some columns, in addition to carrying axial loads, support lateral loads and/or resist moments between the column ends. Such columns are called *beam-columns*. These columns are checked by

$$\left(\frac{f_c}{F'_c}\right)^2 + \frac{f_{bx}}{F'_{bx}\left(1 - \frac{f_c}{F_{cEx}}\right)}$$
$$+ \frac{f_{by}}{F'_{by}\left(1 - \frac{f_c}{F_{cEy}} - \left(\frac{f_{bx}}{F_{bE}}\right)^2\right)} \leq 1.0$$

[NDS Eq. 3.9-3]

F'_c = allowable compressive stress based on the critical value of

$$\left(\frac{l_e}{d}\right)_x \text{ or } \left(\frac{l_e}{d}\right)_y$$
$$= F_c C_D C_M C_t C_F C_i C_P$$
$$= F_c^* C_P$$

F'_{by} = allowable bending stress about y-axis for all sizes of sawn lumber except beams and stringers category
$$= F_{by}C_D C_M C_t C_F C_{fu}C_i \text{ for sawn lumber}$$
$$= F_{by}C_D C_M C_t C_{fu} \text{ for glulam}$$

$F'_{bx} = F_b C_D C_M C_t C_F C_L C_r C_i$ for sawn lumber

C_r (See NDS Supp. Tables 4A, 4B, 4C, and 4E.)

For glulam, F'_{bx} is equal to the smaller of $F'_{bx} = F_{bx}C_D C_M C_t C_L$ or $F'_{bx} = F_{bx}C_D C_M C_t C_V$.

3.7.1
$$F_{cEx} = \frac{K_{cE}E'_x}{\left(\left(\frac{l_e}{d}\right)_x\right)^2}$$

3.7.1
$$F_{cEy} = \frac{K_{cE}E'_y}{\left(\left(\frac{l_e}{d}\right)_y\right)^2}$$

3.3.3.8
$$F_{bE} = \frac{K_{bE}E'_y}{R_B^2}$$

Example 6.7
Combined Compression and Bending: Sawn Lumber

The roof rafter shown is to support a total roof load of 44.1 lbf/ft, which is the combined dead and snow load. Maximum forces and moments are as follows.

moment about x-axis, M_x	1940 ft-lbf
shear, V	391 lbf
axial compression, P	196 lbf

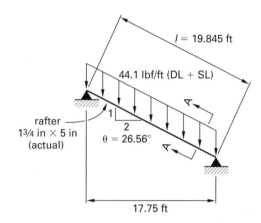

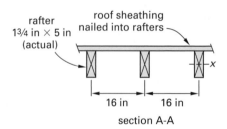

section A-A

The rafter has dressed (actual) dimensions of $1\frac{3}{4}$ in $\times$ 5 in. Assume no. 2 SPF and a wet service condition. Assume roof sheathing provides continuous lateral support to rafter.

Determine if the rafter is adequately sized.

Solution:
NDS Section

For a $1\frac{3}{4}$ in $\times$ 5 in section,

$$A = (1.75 \text{ in})(5 \text{ in}) = 8.75 \text{ in}^2$$

$$I_x = \frac{(1.75 \text{ in})(5 \text{ in})^3}{12} = 18.2 \text{ in}^4$$

$$S_x = \frac{(1.75 \text{ in})(5 \text{ in})^2}{6} = 7.29 \text{ in}^3$$

Supp.
Table 4A;
Adj.
Factors

		C_F	C_r	C_M
$F_b = 875 \text{ lbf/in}^2$		1.4	1.15	0.85
$F_v = 70 \text{ lbf/in}^2$		–	–	0.97
$F_c = 1150 \text{ lbf/in}^2$		1.1	–	0.80
$E = 1,400,000 \text{ lbf/in}^2$		–	–	0.90

Table 2.3.2   For snow load, $C_D = 1.15$.

A. Axial compression

$$f_c = \frac{P}{A} = \frac{196.0 \text{ lbf}}{(1.75 \text{ in})(5 \text{ in})}$$

$$= 22.4 \text{ lbf/in}^2$$

$$F_c' = F_c^* C_P$$

Table 2.3.1 $F_c^* = F_c C_D C_M C_t C_F C_i$

$$= \left(1150 \frac{\text{lbf}}{\text{in}^2}\right)(1.15)(0.80)(1.0)(1.10)(1.0)$$

$$= 1163.8 \text{ lbf/in}^2$$

3.7.1   Find the column stability factor, C_P.

Because of continuous lateral support provided by roof sheathing, assume movement in the direction of the x-axis is prevented. Accordingly,

$$\left(\frac{l_e}{d}\right)_y = 0$$

$$\left(\frac{l_e}{d}\right)_x = \left(\frac{K_e l_u}{d}\right)_x$$

$$= \left(\frac{(1.0)(19.845 \text{ ft})\left(12 \frac{\text{in}}{\text{ft}}\right)}{5 \text{ in}}\right)_x$$

$$= 47.63 = \left(\frac{l_e}{d}\right)_{max}$$

$$E' = E C_M C_t C_i$$

$$= \left(1,400,000 \frac{\text{lbf}}{\text{in}^2}\right)(0.90)(1.0)(1.0)$$

$$= 1,260,000 \text{ lbf/in}^2$$

Eq. 3.7-1   For sawn lumber, $K_{cE} = 0.3$ and $c = 0.8$.

$$F_{cE} = \frac{K_{cE} E'}{\left(\left(\frac{l_e}{d}\right)_{max}\right)^2}$$

$$= \frac{(0.3)\left(1,260,000 \frac{\text{lbf}}{\text{in}^2}\right)}{(47.63)^2}$$

$$= 166.6 \text{ lbf/in}^2$$

$$\frac{F_{cE}}{F_c^*} = \frac{166.6 \frac{\text{lbf}}{\text{in}^2}}{1163.8 \frac{\text{lbf}}{\text{in}^2}} = 0.143$$

$$\frac{1 + \frac{F_{cE}}{F_c^*}}{2c} = \frac{1 + 0.143}{(2)(0.8)} = 0.714$$

$$C_P = \frac{1 + \frac{F_{cE}}{F_c^*}}{2c} - \sqrt{\left(\frac{1 + \frac{F_{cE}}{F_c^*}}{2c}\right)^2 - \frac{\frac{F_{cE}}{F_c^*}}{c}}$$

$$= 0.714 - \sqrt{(0.714)^2 - \frac{0.143}{0.8}}$$

$$= 0.139$$

$$F_c' = F_c^* C_P$$

$$= \left(1163.8 \frac{\text{lbf}}{\text{in}^2}\right)(0.139)$$

$$= 161.8 \text{ lbf/in}^2$$

$$161.8 \text{ lbf/in}^2 > 22.4 \text{ lbf/in}^2$$

$$F_c' > f_c \quad [\text{OK}]$$

B. Bending (ignore M_y)

$$f_{bx} = \frac{M_x}{S_x}$$

$$= \frac{(1940 \text{ ft-lbf})\left(12 \frac{\text{in}}{\text{ft}}\right)}{7.29 \text{ in}^3}$$

$$= 3193.4 \text{ lbf/in}^2$$

$$f_{by} = 0$$

$$F_{bx}' = F_b C_D C_M C_t C_F C_L C_r$$

The rafter has full lateral support. Therefore, $l_u = 0$ (so $l_e = 0$) and $C_L = 1.0$.

$$F_{bx}' = F_b C_D C_M C_t C_F C_L C_r$$

$$= \left(875 \frac{\text{lbf}}{\text{in}^2}\right)(1.15)(0.85)$$

$$\times (1.0)(1.4)(1.0)(1.15)$$

$$= 1377.0 \text{ lbf/in}^2$$

$$1377.0 \text{ lbf/in}^2 < 3193.4 \text{ lbf/in}^2$$

$$F_{bx}' < f_{bx} \quad [\text{no good}]$$

C. Combined compression and bending

Eq. 3.9-3 $$\left(\frac{f_c}{F_c'}\right)^2 + \frac{f_{bx}}{F_{bx}'\left(1 - \frac{f_c}{F_{cEx}}\right)} \leq 1.0$$

$$F_{cEx} = \frac{K_{cE}E'}{\left(\left(\dfrac{l_e}{d}\right)_x\right)^2}$$

$$= \frac{(0.3)\left(1{,}260{,}000 \ \dfrac{\text{lbf}}{\text{in}^2}\right)}{(47.63)^2}$$

$$= 166.6 \ \text{lbf/in}^2$$

$$\left(\frac{22.4 \ \dfrac{\text{lbf}}{\text{in}^2}}{161.8 \ \dfrac{\text{lbf}}{\text{in}^2}}\right)^2 + \frac{3193.4 \ \dfrac{\text{lbf}}{\text{in}^2}}{\left(1377 \ \dfrac{\text{lbf}}{\text{in}^2}\right)\left(1 - \dfrac{22.4 \ \dfrac{\text{lbf}}{\text{in}^2}}{166.6 \ \dfrac{\text{lbf}}{\text{in}^2}}\right)}$$

$$= 2.69 > 1.0 \quad [\text{no good}]$$

Example 6.8
Combined Compression and Bending: Glulam

A glulam bending combination 20F-V2 (SP/SP) of $6^3/_4$ in $\times$ $12^3/_8$ in is laterally supported at its ends and at its midpoint. Assume $C_D = C_t = 1.0$ and MC > 16%.

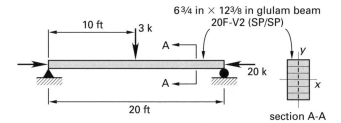

6¾ in × 12⅜ in glulam beam
20F-V2 (SP/SP)

section A-A

Check the adequacy of the member.

Solution:

NDS Section

Supp.
Table 1C

For a $6^3/_4$ in $\times$ $12^3/_8$ in, nine lams, $A = 83.53$ in^2 and $S_x = 172.3$ in^3.

Supp.
Table 5A;
Adj. Factors

		C_M
$F_{bx} = 2000 \ \text{lbf/in}^2$		0.80 $K_L = 1.09$ (for C_V)
$F_{by} = 1450 \ \text{lbf/in}^2$		0.80
$E_x = 1{,}600{,}000 \ \text{lbf/in}^2$		0.833
$E_y = 1{,}400{,}000 \ \text{lbf/in}^2$		0.833
$E = 1{,}400{,}000 \ \text{lbf/in}^2$		0.833
$F_{vx} = 240 \ \text{lbf/in}^2$		0.875
$F_{vy} = 210 \ \text{lbf/in}^2$		0.875
$F_c = 1550 \ \text{lbf/in}^2$		0.73

A. Axial compression

$$f_c = \frac{P}{A} = \frac{20{,}000 \ \text{lbf}}{83.53 \ \text{in}^2}$$

$$= 239.4 \ \text{lbf/in}^2$$

$$F_c' = F_c^* C_P$$

Table 2.3.1

$$F_c^* = F_c C_D C_M C_t$$

$$= \left(1550 \ \frac{\text{lbf}}{\text{in}^2}\right)(1.0)(0.73)(1.0)$$

$$= 1131.5 \ \text{lbf/in}^2$$

3.7.1

Find the column stability factor, C_P.

Eq. 3.7-1

$$\left(\frac{l_e}{d}\right)_x = \left(\frac{K_e l_u}{d}\right)_x$$

$$= \frac{(1.0)(20 \ \text{ft})\left(12 \ \dfrac{\text{in}}{\text{ft}}\right)}{12.375 \ \text{in}}$$

$$= 19.39 = \left(\frac{l_e}{d}\right)_{\text{max}}$$

$$\left(\frac{l_e}{d}\right)_y = \left(\frac{K_e l_u}{d}\right)_y$$

$$= \frac{(1.0)(10 \ \text{ft})\left(12 \ \dfrac{\text{in}}{\text{ft}}\right)}{6.75 \ \text{in}}$$

$$= 17.8$$

Therefore, the x-axis (the strong axis) of the beam-column is critical and $(l_e/d)_x$ is used to determine F_c'.

$$E_x' = E_x C_M C_t$$

$$= \left(1{,}600{,}000 \ \frac{\text{lbf}}{\text{in}^2}\right)(0.833)(1.0)$$

$$= 1{,}332{,}800 \ \text{lbf/in}^2$$

3.7.1.5

For glulam, $K_{cE} = 0.418$ and $c = 0.9$.

$$F_{cEx} = \frac{K_{cEx}E_x'}{\left(\left(\dfrac{l_e}{d}\right)_x\right)^2}$$

$$= \frac{(0.418)\left(1{,}332{,}800 \ \dfrac{\text{lbf}}{\text{in}^2}\right)}{(19.39)^2}$$

$$= 1481.8 \ \text{lbf/in}^2$$

$$\frac{F_{cEx}}{F_c^*} = \frac{1481.8 \ \dfrac{\text{lbf}}{\text{in}^2}}{1131.5 \ \dfrac{\text{lbf}}{\text{in}^2}} = 1.31$$

$$\frac{1 + \dfrac{F_{cEx}}{F_c^*}}{2c} = \frac{1 + 1.31}{(2)(0.9)} = 1.283$$

$$C_P = \frac{1 + \dfrac{F_{cEx}}{F_c^*}}{2c} - \sqrt{\left(\frac{1 + \dfrac{F_{cEx}}{F_c^*}}{2c}\right)^2 - \frac{\dfrac{F_{cEx}}{F_c^*}}{c}}$$

$$= 1.283 - \sqrt{(1.283)^2 - \frac{1.31}{0.9}}$$

$$= 0.846$$

$$F_c' = F_c^* C_P$$

$$= \left(1131.5 \ \frac{\text{lbf}}{\text{in}^2}\right)(0.846)$$

$$= 957.2 \ \text{lbf/in}^2$$

$$957.2 \ \frac{\text{lbf}}{\text{in}^2} > 239.4 \ \text{lbf/in}^2$$

$$F_c' > f_c \quad [\text{OK}]$$

B. Bending (about x-axis only)

$$M_x = \frac{PL}{4}$$

$$= \frac{(3000 \ \text{lbf})(20 \ \text{ft})\left(12 \ \dfrac{\text{in}}{\text{ft}}\right)}{4}$$

$$= 180{,}000 \ \text{in-lbf}$$

$$f_{bx} = \frac{M_x}{S_x}$$

$$= \frac{180{,}000 \ \text{ft-lbf}}{172.3 \ \text{in}^3}$$

$$= 1044.7 \ \text{lbf/in}^2$$

Table 2.3.1
Ftn. 1   For glulam timber bending, the smaller of the two values, C_L and C_V, controls the following: $F_{bx}' = F_{bx}C_D C_M C_t C_L = F_{bx}^* C_L$ and $F_{bx}' = F_{bx}C_D C_M C_t C_V = F_{bx}^* C_V$.

3.3.3   Find the beam stability factor, C_L.

Eq. 3.3-6   Lateral-torsional buckling will occur in the plane of the y-axis.

For the unbraced length,

$$l_u = (10 \ \text{ft})\left(12 \ \frac{\text{in}}{\text{ft}}\right) = 120 \ \text{in}$$

Table 3.3.3   For a concentrated load at center, the effective unbraced length is
$$l_e = 1.11 l_u = (1.11)(120 \ \text{in}) = 133.2 \ \text{in}$$

$$R_B = \sqrt{\frac{l_e d}{b^2}}$$

$$= \sqrt{\frac{(133.2 \ \text{in})(12.375 \ \text{in})}{(6.75 \ \text{in})^2}}$$

$$= 6.02$$

$$E_y' = E_y C_M C_t$$

$$= \left(1{,}400{,}000 \ \frac{\text{lbf}}{\text{in}^2}\right)(0.833)(1.0)$$

$$= 1{,}166{,}200 \ \text{lbf/in}^2$$

3.3.3.8   For glulam, $K_{bE} = 0.610$.

$$F_{bE} = \frac{K_{bE} E_y'}{R_B^2}$$

$$= \frac{(0.610)\left(1{,}166{,}200 \ \dfrac{\text{lbf}}{\text{in}^2}\right)}{(6.02)^2}$$

$$= 19{,}629.6 \ \text{lbf/in}^2$$

$$F_{bx}^* = F_{bx} C_D C_M C_t$$

$$= \left(2000 \ \frac{\text{lbf}}{\text{in}^2}\right)(1.0)(0.833)(1.0)$$

$$= 1666 \ \text{lbf/in}^2$$

$$\frac{F_{bE}}{F_{bx}^*} = \frac{19{,}629.6 \ \dfrac{\text{lbf}}{\text{in}^2}}{1666 \ \dfrac{\text{lbf}}{\text{in}^2}} = 11.8$$

$$\frac{1 + \dfrac{F_{bE}}{F_{bx}^*}}{1.9} = \frac{1 + 11.8}{1.9} = 6.74$$

Eq. 3.3-6
$$C_L = \frac{1 + \dfrac{F_{bE}}{F_{bx}^*}}{1.9} - \sqrt{\left(\frac{1 + \dfrac{F_{bE}}{F_{bx}^*}}{1.9}\right)^2 - \frac{\dfrac{F_{bE}}{F_{bx}^*}}{0.95}}$$

$$= 6.74 - \sqrt{(6.74)^2 - \frac{11.8}{0.95}}$$

$$= 0.99 \leq 1.0$$

5.3.2;
Supp.
Table 5A   Find the volume factor of glulam, C_V.

Eq. 5.3-1 $$C_V = K_L \left(\frac{21}{L}\right)^{1/x} \left(\frac{12}{d}\right)^{1/x} \left(\frac{5.125}{b}\right)^{1/x}$$

Table 5.3.2 $K_L = 1.09$

For southern pine, $x = 20$.

$$C_V = (1.09) \left(\frac{21}{20 \text{ ft}} \right)^{1/20} \left(\frac{12}{12.375 \text{ in}} \right)^{1/20}$$

$$\times \left(\frac{5.125}{6.75 \text{ in}} \right)^{1/20}$$

$$= 1.08 \geq 1.0$$

Therefore, let $C_V = 1.0$.

$1.0 > 0.99$
$C_V > C_L$

Therefore, C_L governs lateral stability.

$$F'_{bx} = F^*_{bx} C_L$$

$$= \left(1666 \ \frac{\text{lbf}}{\text{in}^2} \right) (0.99)$$

$$= 1649.3 \ \text{lbf/in}^2$$

C. Combined compression and bending

Eq. 3.9-3 $$\left(\frac{f_c}{F'_c} \right)^2 + \frac{f_{bx}}{F'_{bx} \left(1 - \dfrac{f_c}{F_{cEx}} \right)} \leq 1.0$$

$$\left(\frac{239.4 \ \dfrac{\text{lbf}}{\text{in}^2}}{957.2 \ \dfrac{\text{lbf}}{\text{in}^2}} \right)^2 + \frac{1044.7 \ \dfrac{\text{lbf}}{\text{in}^2}}{\left(1649.3 \ \dfrac{\text{lbf}}{\text{in}^2} \right) \left(1 - \dfrac{239.4 \ \dfrac{\text{lbf}}{\text{in}^2}}{1481.8 \ \dfrac{\text{lbf}}{\text{in}^2}} \right)}$$

$$= 0.818 < 1.0 \quad [\text{OK}]$$

Mechanical Connections

Parts 7 to 14 of the 1997 NDS cover the design of mechanical connections, bolts, lag screws/bolts, wood screws, nails and spikes, and other connectors. Nominal design values for laterally loaded bolts, lag screws, wood screws, and nails and spikes are based on yield limit equations, which model the various yield modes (see NDS App. I). These design values in a given species apply to all grades unless otherwise noted.

1. Typical Lateral (Shear) Connections

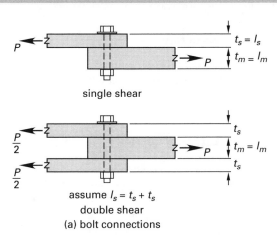

single shear

assume $l_s = t_s + t_s$
double shear
(a) bolt connections

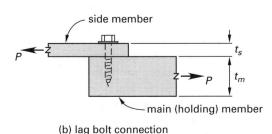

(b) lag bolt connection

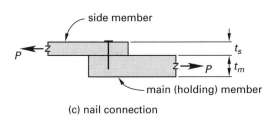

(c) nail connection

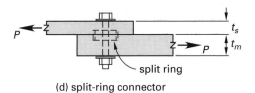

(d) split-ring connector

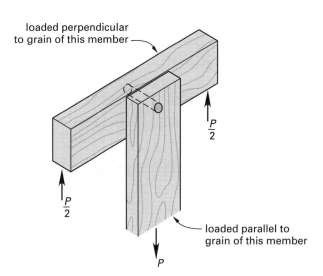

(e) load relative to grain direction

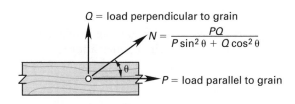

Q = load perpendicular to grain

$$N = \frac{PQ}{P \sin^2 \theta + Q \cos^2 \theta}$$

P = load parallel to grain

(f) N, inclined load at angle θ

Figure 7.1 Mechanical Connectors: Shear

2. Typical Tension (Withdrawal) Connections

Withdrawal refers to the pulling out of a connector from the wood as caused by tensile forces.

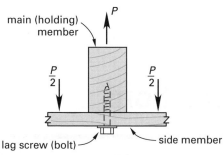

(a) side grain loads (perpendicular to grain)

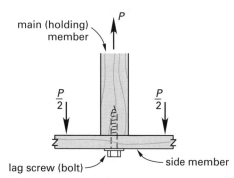

(b) end grain loads (parallel to grain, not allowed for nails and wood screws)

Figure 7.2 Mechanical Connectors: Withdrawal

3. Adjustment Factors

[NDS 7.3, NDS Tables 7.3.1 and 7.3.3]

Nominal (tabulated) design values, Z and W, are multiplied by all applicable adjustment factors (NDS 7.3) to determine allowable design values, Z' and W' (see Tables 7.1 and 7.2).

Z = nominal lateral design value for a single fastener connection

W = nominal withdrawal design value for a fastener per inch of penetration

Z' = allowable lateral design value for a single fastener connection

W' = allowable withdrawal design value for a fastener per inch of penetration

A. Load Duration Factor, C_D

[NDS Table 2.3.2, NDS App. B, NDS 7.3.2]

C_D values used previously for beams and columns are also applicable to the wood parts of the connections except when the connector capacity is controlled by the metal parts of the connections (plates, hangers, fasteners, etc.). Furthermore, the impact load duration factor is not applicable to connections.

B. Wet Service Factor, C_M

[NDS Table 7.3.3]

Nominal (tabulated) design values are for connections in wood with a moisture content of 19% or less, at the time of fabrication *and* in service, and are multiplied by the wet service factors, C_M, when different conditions are encountered (see Table 7.2).

C. Group Action Factor, C_g

[NDS 7.3.6; Eq. 7.3-1; NDS Tables 7.3.6A–7.3.6D]

Nominal (tabulated) design values are multiplied by group action factors, C_g, specified in NDS 7.3.6.

NDS Eq. 7.3-1 and NDS Tables 7.3.6A through 7.3.6D provide C_g values for various connection geometries.

D. Other Adjustment Factors

Adjustment factors, such as connection geometry, C_Δ, penetration factor, C_d, and so on, will be covered as they are needed.

4. Laterally Loaded Fasteners

The capacities of laterally loaded nails, bolts, and lag screws/bolts were based on empirical work until publication of the 1997 NDS. The 1997 NDS, however, used an engineering mechanics approach based on yield limit theory for analyzing dowel-type fasteners in wood connections, resulting in applicable equations. NDS App. I provides more detail on the failure mechanisms considered for development of the yield limit equations for connections. In addition to these equations for laterally loaded nails, bolts, and lag screws/bolts, the 1997 NDS also provides a limited set of load tables for common, limited conditions (as noted in the footnotes to each table). While these tables are based on the yield limit equations, they have been simplified by grouping parameters together.

Table 7.1 Applicability of Adjustment Factors for Connections (NDS Table 7.3.1)

		load duration factor [a]	wet service factor [b]	temperature factor	group action factor	geometry factor [c]	penetration depth factor [c]	end grain factor [c]	metal side plate factor [c]	diaphragm factor [c]	toe-nail factor [c]
bolts	$Z' = (Z)$	(C_D)	(C_M)	(C_t)	(C_g)	(C_Δ)	–	–	–	–	–
lag screws	$W' = (W)$	(C_D)	(C_M)	(C_t)	–	–	–	(C_{eg})	–	–	–
	$Z' = (Z)$	(C_D)	(C_M)	(C_t)	(C_g)	(C_Δ)	(C_d)	(C_{eg})	–	–	–
split ring and shear plate connectors	$P' = (P)$	(C_D)	(C_M)	(C_t)	(C_g)	(C_Δ)	(C_d)	–	(C_{st})	–	–
	$Q' = (Q)$	(C_D)	(C_M)	(C_t)	(C_g)	(C_Δ)	(C_d)	–	–	–	–
wood screws	$W' = (W)$	(C_D)	(C_M)	(C_t)	–	–	–	–	–	–	–
	$Z' = (Z)$	(C_D)	(C_M)	(C_t)	–	–	(C_d)	(C_{eg})	–	–	–
nails and spikes	$W' = (W)$	(C_D)	(C_M)	(C_t)	–	–	–	–	–	–	(C_{tn})
	$Z' = (Z)$	(C_D)	(C_M)	(C_t)	–	–	(C_d)	(C_{eg})	–	(C_{di})	(C_{tn})
metal plate connectors	$Z' = (Z)$	(C_D)	(C_M)	(C_t)	–	–	–	–	–	–	–
drift bolts and drift pins	$W' = (W)$	(C_D)	(C_M)	(C_t)	–	–	–	(C_{eg})	–	–	–
	$Z' = (Z)$	(C_D)	(C_M)	(C_t)	(C_g)	(C_Δ)	(C_d)	(C_{eg})	–	–	–
spike grids	$Z' = (Z)$	(C_D)	(C_M)	(C_t)	–	(C_Δ)	–	–	–	–	–
timber rivets	$P' = (P)$	$(C_D{}^d)$	(C_M)	(C_t)	–	–	–	–	$(C_{st}{}^e)$	–	–
	$Q' = (Q)$	$(C_D{}^d)$	(C_M)	(C_t)	–	$(C_\Delta{}^f)$	–	–	$(C_{st}{}^e)$	–	–

[a] The load duration factor, C_D, shall not exceed 1.6 for connections (NDS 7.3.2).

[b] The wet service factor, C_M, shall not apply to toe-nails loaded in withdrawal (NDS 12.2.3).

[c] Specific information concerning geometry factors (C_Δ), penetration depth factors (C_d), end grain factors (C_{eg}), metal side plate factors (C_{st}), diaphragm factors (C_{di}), and toe-nail factors (C_{tn}) is provided in NDS Chs. 8, 9, 10, 11, 12, and 14.

[d] The load duration factor, C_D, is only applied when wood capacity (P_w, Q_w) controls. See NDS Ch. 13 for specific information.

[e] The metal side plate factor, C_{st}, is only applied when rivet capacity (P_r, Q_r) controls. See NDS Ch. 13 for specific information.

[f] The geometry factor, C_Δ, is only applied when wood capacity, Q_w, controls. See NDS Ch. 13 for specific information.

Table 7.2 Wet Service Factors, C_M, for Connections (NDS Table 7.3.3)[a]

fastener type	moisture content		load	
	at time of fabrication	in-service	lateral	withdrawal
shear plates split rings[b]	≤19%	≤19%	1.0	–
	>19%	≤19%	0.8	–
	any	>19%	0.7	–
metal connector plates[c]	≤19%	≤19%	1.0	–
	>19%	≤19%	0.8	–
	any	>19%	0.8	–
bolts and drift pins and drift bolts	≤19%	≤19%	1.0	–
	>19%	≤19%	0.4[d]	–
	any	>19%	0.7	
lag screws and wood screws	≤19%	≤19%	1.0	1.0
	>19%	≤19%	0.4[d]	1.0
	any	>19%	0.7	0.7
nails and spikes	≤19%	≤19%	1.0	1.0
	>19%	≤19%	0.7	0.25
	≤19%	>19%	0.7	0.25
	>19%	>19%	0.7	1.0
threaded hardened nails	any	any	1.0	1.0
timber rivets	≤19%	≤19%	1.0	–
	>19%	≤19%	0.9	–
	≤19%	>19%	0.8	–
	>19%	>19%	0.8	–

[a] 1998 Identified Errata to 1997 Edition of NDS-1997 published September 1998 is incorporated.

[b] For split-ring or shear-plate connectors, moisture content limitations apply to a depth of $^3/_4$ in below the surface of the wood.

[c] For more information on metal connector plates, see NDS 14.3.

[d] $C_M = 1.0$ for wood screws. For bolt and lag screw connections with: 1) one fastener only, or 2) two or more fasteners placed in a single row parallel to grain, or 3) fasteners placed in two or more rows parallel to grain with separate splice plates for each row, $C_M = 1.0$.

Nails and Spikes

There are four basic types of nails: box nails, common wire nails, common wire spikes, and threaded hardened-steel nails. For a given pennyweight, all four have the same length. The first three nails have the same form but different diameters: box nails have the smallest diameter, and common wire spikes have the largest diameter. The threaded hardened-steel nails are made from high-carbon steel and have an annularly, or helically, threaded shank, which provides better withdrawal capacity than other nails.

The sizes of these nails are given in NDS Tables 12.3A through 12.3H (which also list lateral, i.e., shear, design values). Nail diameters vary from $3/32$ in to $3/8$ in, and their lengths can be from 2 in to 9 in depending upon their size classifications.

1. Withdrawal Design Values
[NDS 12.2]

A. Nominal (Tabulated) Withdrawal Design Value, W
[NDS Tables 12A and 12.2A]

Nominal (tabulated) withdrawal design values, W, in lbf per inch of penetration, for a single nail or spike driven in the side grain of the main member, depend upon the specific gravities of the wood species (NDS Table 12A) and are given in NDS Table 12.2A.

Nails and spikes are not allowed to be loaded in withdrawal from the end grain of the wood.

B. Toe-Nail Factor, C_{tn}
[NDS 12.2.3]

The toe-nail factor, C_{tn}, to be multiplied to the nominal design values is 0.67. But the wet service factor, C_M, does not apply to toe-nails loaded in withdrawal.

C. Allowable Withdrawal Design Value, W'
[NDS Table 7.3.1]

The allowable design value, W', in lbf per inch of penetration into the main member depends on the nominal value, W, obtained from the table.

$$W' = WpC_D C_M C_t C_{tn}$$

$p =$ distance of penetration of nail into the holding (main) member (in)

Example 8.1
Withdrawal Capacity of Nails

Two 16d spikes connect 2×6 to 4×8 hem-fir lumbers as shown. The connected lumbers will be subjected to wetting in service. Conditions are subjected to normal temperatures ($C_t = 1.0$).

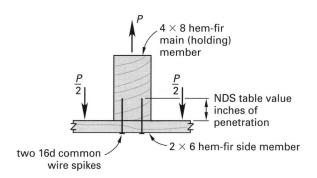

Determine the allowable withdrawal value of two 16d spikes for combined snow and dead loads.

Solution:

NDS Section

Table 2.3.2   For snow load, $C_D = 1.15$.

Table 7.3.3 $C_M = 0.25$

Table 12A   For hem-fir, $G = 0.43$.

Table 12.3D   For a 16d spike,

           diameter, D 0.207 in
           length, L 3.5 in

Table 12.2A   For the given specifc gravity, G, and spike diameter, D, $W = 35$ lbf per inch of penetration into main member.

Actual penetration, p, into a 4×8 holding member is

$$p = 3.5 \text{ in} - 1.5 \text{ in}$$
$$= 2.0 \text{ in} \quad \begin{bmatrix} \text{dressed depth for a} \\ 2 \times 6 \text{ board is } 1.5 \text{ in} \end{bmatrix}$$

Table 7.3.1
$$W' = WpC_DC_MC_t$$

$$= \left(35 \frac{\text{lbf}}{\text{in}}\right)(2.0 \text{ in})(1.15)(0.25)(1.0)$$

$$= 20.1 \text{ lbf per spike}$$

$$\left(20.1 \frac{\text{lbf}}{\text{spike}}\right)(2 \text{ spikes}) = 40.2 \text{ lbf}$$

2. Lateral Design Values
[NDS 12.3; Eqs. 12.3-1−12.3-4; Tables 12.3A−12.3H]

A. Nominal (Tabulated) Lateral Design Value, Z

The nominal lateral design values, Z, are the lesser of those determined by NDS yield limit Eqs. 12.3-1 through 12.3-4.

For Wood-to-Wood Connections
[NDS 12.3.1]

$$Z = \frac{Dt_sF_{es}}{K_D} \qquad \text{[NDS Eq. 12.3-1]}$$

$$Z = \frac{k_1DpF_{em}}{K_D(1+2R_e)} \qquad \text{[NDS Eq. 12.3-2]}$$

$$Z = \frac{k_2Dt_sF_{em}}{K_D(2+R_e)} \qquad \text{[NDS Eq. 12.3-3]}$$

$$Z = \frac{D^2}{K_D}\sqrt{\frac{2F_{em}F_{yb}}{(3)(1+R_e)}} \qquad \text{[NDS Eq. 12.3-4]}$$

$$k_1 = -1 + \sqrt{(2)(1+R_e) + \frac{2F_{yb}(1+2R_e)D^2}{3F_{em}p^2}}$$

$$k_2 = -1 + \sqrt{\frac{(2)(1+R_e)}{R_e} + \frac{2F_{yb}(2+R_e)D^2}{3F_{em}t_s^2}}$$

$$R_e = \frac{F_{em}}{F_{es}}$$

p = penetration of nail or spike into holding (main) member (in)

$F_{em} = 16{,}600G^{1.84}$ [see NDS Table 12A]

G (See NDS Tables 8A, 11A, and 12A.)

F_{es} (See NDS Table 12A.)

F_{yb} (See NDS Table 12.3A Ftns.)

D = nail or spike diameter (when annularly threaded nails are used with threads at the shear plane, D = root diameter of threaded portion of nail)

$K_D = 2.2$ for $D \leq 0.17$ in
$K_D = 10D + 0.5$ for $0.17 \text{ in} < D < 0.25$ in
$K_D = 3.0$ for $D \geq 0.25$ in

For Wood-to-Metal Connections
[NDS 12.3.2]

NDS Eqs. 12.3-2, 12.3-3, and 12.3-4 are also used with F_{es} = dowel bearing strength of the metal side member, ASTM A446, Grade A steel ($F_{es} = 45{,}000$ lbf/in^2).

As an alternative to these equations, NDS Tables 12.3A through 12.3H contain nominal design values, Z, for a set of common cases only. For those cases not contained in these tables, the yield limit equations must be used.

The Z values given by equations or tables are for such conditions as nails driven into the side grain of the holding member; penetration length, p, into the holding member of $12D$; 10-year load duration; dry at time of fabrication and in service; normal temperature; and other relevant conditions.

B. Allowable Lateral Design Value, Z', and Adjustment Factors
[NDS Table 7.3.1; NDS 12.3.4 to 12.3.7]

The allowable lateral design values, Z', are determined by multiplying all applicable adjustment factors (NDS Table 7.3.1) to the nominal lateral design values.

$$Z' = ZC_DC_MC_tC_dC_{eg}C_{di}C_{tn}$$

$C_d = p/12D \leq 1.0$, where $6D \leq p \leq 12D$ (See NDS 12.3.4 and NDS Eq. 12.3-5.)

$C_{eg} = 0.67$ when driven in the end grain (See NDS 12.3.5.)

$C_{di} = 1.1$ when used in a diaphragm construction (See NDS 12.3.6.)

$C_{tn} = 0.83$ (Note: different from withdrawal toenail factor, which is 0.67) (See NDS 12.3.7.)

Example 8.2
Lateral Design Capacity by NDS Table 12.3 (A−H): Tension Splice

Six 40d common wire spikes hold two members, 2×6 and 4×8, together as shown. Both members are douglas fir-larch under wet service. The loads are DL + SL. Assume $C_t = 1.0$.

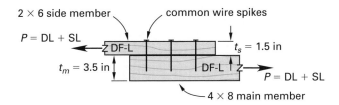

2 × 6 side member — common wire spikes

$P = DL + SL$

DF-L

$t_s = 1.5$ in

$t_m = 3.5$ in

DF-L

$P = DL + SL$

4 × 8 main member

Determine the allowable lateral design load for the six 40d common wire spikes using NDS Table 12.3D.

Solution:

NDS Section

Table 12.3D For douglas fir-larch, $t_s = 1.5$ in.

For 40d spike, $L = 5.0$ in; $D = 0.263$ in.

The nominal design value is $Z = 240$ lbf for douglas fir-larch.

Table 2.3.2 For snow load, $C_D = 1.15$.

Table 7.3.3 For wet service connection, $C_M = 0.70$.

12.3.4 $p = 5.0$ in $- 1.5$ in $= 3.5$ in

Eq. 12.3-5 $C_d = \dfrac{p}{12D} = \dfrac{3.5 \text{ in}}{(12)(0.263 \text{ in})}$

$= 1.11 > 1.0$

Therefore, $C_d = 1.0$.

Table 7.3.1 The allowable lateral design value is
$$Z' = ZC_DC_MC_tC_d$$
$$= (240 \text{ lbf})(1.15)(0.70)(1.0)(1.0)$$
$$= 193.2 \text{ lbf}$$

$$\left(193.2 \; \frac{\text{lbf}}{\text{spike}}\right)(6 \text{ spikes}) = 1159.2 \text{ lbf}$$

Example 8.3
Lateral Design Capacity by
NDS Equation 12.3 (1−4): Tension Splice

For the configuration shown in Ex. 8.2, determine the nominal lateral design value, Z, using NDS Eqs. 12.3-1 through 12.3-4.

Solution:

NDS Section

12.3.1 $D = 0.263$ in; $p = 3.5$ in.

Therefore, $K_D = 3.0$ since $D \geq 0.25$ in.

Table 12A For douglas fir-larch, $F_{em} = F_{es} = 4650$ lbf/in².

12.3.1; $t_s = 1.5$ in

Table 12.3D
Ftn. For a 0.263 in diameter spike, $F_{yb} = 70,000$ lbf/in².

Eq. 12.3-1 $Z = \dfrac{Dt_sF_{es}}{K_D}$

$$= \dfrac{(0.263 \text{ in})(1.5 \text{ in})\left(4650 \; \dfrac{\text{lbf}}{\text{in}^2}\right)}{3.0}$$

$$= 611.5 \text{ lbf}$$

Table 12A For douglas fir-larch, $F_e = 4650.0$ lbf/in².

Eq. 12.3-2 $R_e = \dfrac{F_{em}}{F_{es}} = \dfrac{4650 \; \dfrac{\text{lbf}}{\text{in}^2}}{4650 \; \dfrac{\text{lbf}}{\text{in}^2}} = 1.0$

$k_1 = -1 + \sqrt{(2)(1 + R_e) + \dfrac{2F_{yb}(1 + 2R_e)D^2}{3F_{em}p^2}}$

$$= -1 + \sqrt{(2)(1 + 1.0) + \dfrac{(2)\left(70,000 \; \dfrac{\text{lbf}}{\text{in}^2}\right) \times (1 + (2)(1.0))(0.263 \text{ in})^2}{(3)\left(4650 \; \dfrac{\text{lbf}}{\text{in}^2}\right)(3.5 \text{ in})^2}}$$

$$= 1.04$$

$Z = \dfrac{k_1DpF_{em}}{K_D(1 + 2R_e)}$

$$= \dfrac{(1.04)(0.263 \text{ in})(3.5 \text{ in})\left(4650 \; \dfrac{\text{lbf}}{\text{in}^2}\right)}{(3)\left(1 + (2)(1.0)\right)}$$

$$= 494.6 \text{ lbf}$$

Eq. 12.3-3

$k_2 = -1 + \sqrt{\dfrac{(2)(1 + R_e)}{R_e} + \dfrac{2F_{yb}(2 + R_e)D^2}{3F_{em}t_s^2}}$

$$= -1 + \sqrt{\dfrac{(2)(1 + 1.0)}{1.0} + \dfrac{(2)\left(70,000 \; \dfrac{\text{lbf}}{\text{in}^2}\right) \times (2 + 1.0)(0.263 \text{ in})^2}{(3)\left(4650 \; \dfrac{\text{lbf}}{\text{in}^2}\right)(1.5 \text{ in})^2}}$$

$$= 1.22$$

$Z = \dfrac{k_2Dt_sF_{em}}{K_D(2 + R_e)}$

$$= \dfrac{(1.22)(0.263 \text{ in})(1.5 \text{ in})\left(4650 \; \dfrac{\text{lbf}}{\text{in}^2}\right)}{(3.0)(2 + 1.0)}$$

$$= 248.7 \text{ lbf}$$

Eq. 12.3-4

$$Z = \frac{D^2}{K_D} \sqrt{\frac{2F_{em}F_{yb}}{(3)(1 + R_e)}}$$

$$= \frac{(0.263 \text{ in})^2}{3.0} \sqrt{\frac{(2)\left(4650 \; \dfrac{\text{lbf}}{\text{in}^2}\right)\left(70{,}000 \; \dfrac{\text{lbf}}{\text{in}^2}\right)}{(3)(1 + 1.0)}}$$

$$= 240.2 \text{ lbf}$$

The nominal lateral design value, Z, is the smallest value given by the four equations. Therefore, $Z = 240.2$ lbf (as compared with $Z = 240$ lbf as determined by NDS Table 12.3D).

Bolts

Bolts, which are dowel-type connections, are used for laterally loaded joints. Connections can be single-shear (two-member), double-shear (three-member), or multiple-shear (several-member) joints. They can also be used for wood-to-wood or wood-to-metal connections.

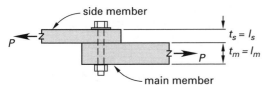

(a) single-shear connection

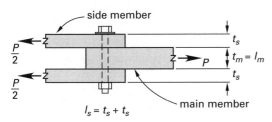

(b) double-shear connection

Figure 9.1 Bolt Connections

Since the yield limit equations for the lateral capacity of bolts in single shear require the dowel bearing strength of wood, numerical values are listed in NDS Table 8A for bolts and NDS Table 9A for lag bolts. These values are based on the following equations.

$$F_{e\parallel} = 11{,}200G$$

$$F_{e\perp} = \frac{6100G^{1.45}}{\sqrt{D}}$$

$F_{e\parallel} =$ nominal bearing strength of wood parallel to grain (lbf/in^2)

$F_{e\perp} =$ nominal bearing strength of wood perpendicular to grain (lbf/in^2)

$D =$ bolt diameter (in)

The specific gravities, G, for different wood species are also listed in NDS Tables 9A, 11A, and 12A.

1. Nominal Design Values for Single-Shear, Two-Member Connections

[NDS 8.2; Eqs. 8.2-1 to 8.2-6; NDS Tables 8.2A–8.2D]

A. Wood-to-Wood Connection

The nominal design value, Z, for one bolt in a single-shear connection is the smallest load capacity obtained from the following six equations.

$$Z = \frac{Dt_m F_{em}}{4K_\theta} \qquad \text{[NDS Eq. 8.2-1]}$$

$$Z = \frac{Dt_s F_{es}}{4K_\theta} \qquad \text{[NDS Eq. 8.2-2]}$$

$$Z = \frac{k_1 Dt_s F_{es}}{3.6K_\theta} \qquad \text{[NDS Eq. 8.2-3]}$$

$$Z = \frac{k_2 Dt_m F_{em}}{(3.2)(1 + 2R_e)K_\theta} \qquad \text{[NDS Eq. 8.2-4]}$$

$$Z = \frac{k_3 Dt_s F_{em}}{(3.2)(2 + R_e)K_\theta} \qquad \text{[NDS Eq. 8.2-5]}$$

$$Z = \frac{D^2}{3.2K_\theta}\sqrt{\frac{2F_{em}F_{yb}}{(3)(1 + R_e)}} \qquad \text{[NDS Eq. 8.2-6]}$$

$Z =$ nominal design value for bolt in single shear (lbf) (Z is to be taken as the smallest value from the six yield limit equations.)

$$k_1 = \frac{\sqrt{\begin{array}{c}R_e + 2R_e^2(1 + R_t + R_t^2) \\ + R_t^2 R_e^3\end{array}} - R_e(1 + R_t)}{1 + R_e}$$

$$k_2 = -1 + \sqrt{(2)(1 + R_e) + \frac{2F_{yb}(1 + 2R_e)D^2}{3F_{em}t_m^2}}$$

$$k_3 = -1 + \sqrt{\frac{(2)(1 + R_e)}{R_e} + \frac{2F_{yb}(2 + R_e)D^2}{3F_{em}t_s^2}}$$

$$R_e = \frac{F_{em}}{F_{es}}$$

$$R_t = \frac{t_m}{t_s}$$

$$K_\theta = 1 + \frac{\theta}{360°}$$

F_e = dowel bearing strength

$F_{em} = F_{e\parallel}$ for load parallel to grain of main member = $11{,}200G$ or NDS Table 8A

$\phantom{F_{em}} = F_{e\perp}$ for load perpendicular to grain = $6100G^{1.45}/\sqrt{D}$ or NDS Table 8A

$\phantom{F_{em}} = F_{e\theta}$ for load at angle to grain θ (See the Hankinson formula.)

$F_{es} = F_{e\parallel}$ for load parallel to grain of side member = $11{,}200G$ or NDS Table 8A

$\phantom{F_{es}} = F_{e\perp}$ for load perpendicular to grain = $6100G^{1.45}/\sqrt{D}$ or NDS Table 8A

$\phantom{F_{es}} = F_{e\theta}$ for load at angle to grain θ (See the Hankinson formula.)

$\phantom{F_{es}}$ For a steel side member, $F_{es} = F_u$.

F_{yb} = bending yield strength of bolt (lbf/in^2)

θ = maximum angle of load to grain; $0 \le \theta \le 90°$ for any member in connection

The dowel bearing strength at an angle of load to grain θ is given by the Hankinson formula.

$$F_{e\theta} = \frac{F_{e\parallel}F_{e\perp}}{F_{e\parallel}\sin^2\theta + F_{e\perp}\cos^2\theta}$$

B. Wood-to-Metal Connection

The nominal design value for one bolt in a single-shear connection between a wood main member and a steel side plate is the smallest load capacity obtained from the following five equations.

$$Z = \frac{Dt_m F_{em}}{4K_\theta} \qquad \text{[NDS Eq. 8.2-1]}$$

$$Z = \frac{k_1 Dt_s F_{es}}{3.6K_\theta} \qquad \text{[NDS Eq. 8.2-3]}$$

$$Z = \frac{k_2 Dt_m F_{em}}{(3.2)(1 + 2R_e)K_\theta} \qquad \text{[NDS Eq. 8.2-4]}$$

$$Z = \frac{k_3 Dt_s F_{em}}{(3.2)(2 + R_e)K_\theta} \qquad \text{[NDS Eq. 8.2-5]}$$

$$Z = \frac{D^2}{3.2K_\theta}\sqrt{\frac{2F_{em}F_{yb}}{(3)(1 + R_e)}} \qquad \text{[NDS Eq. 8.2-6]}$$

Z = nominal design value for bolt in single shear (lbf) (Z is to be taken as the smallest value from the five yield limit equations.)

$$k_1 = \frac{\sqrt{R_e + 2R_e^2(1 + R_t + R_t^2) + R_t^2 R_e^3} - R_e(1 + R_t)}{1 + R_e}$$

$$k_2 = -1 + \sqrt{(2)(1 + R_e) + \frac{2F_{yb}(1 + 2R_e)D^2}{3F_{em}t_m^2}}$$

$$k_3 = -1 + \sqrt{\frac{(2)(1 + R_e)}{R_e} + \frac{2F_{yb}(2 + R_e)D^2}{3F_{em}t_s^2}}$$

$$R_e = \frac{F_{em}}{F_{es}}$$

$$R_t = \frac{t_m}{t_s}$$

$$K_\theta = 1 + \frac{\theta}{360°}$$

$F_{em} = F_{e\parallel}$ for load parallel to grain = $11{,}200G$ or NDS Table 8A

$\phantom{F_{em}} = F_{e\perp}$ for load perpendicular to grain = $6100G^{1.45}/\sqrt{D}$ or NDS Table 8A

$\phantom{F_{em}} = F_{e\theta}$ for load at angle to grain θ (See the Hankinson formula.)

$F_{es} = F_{e\parallel}$ for load parallel to grain = $11{,}200G$ or NDS Table 8A

$\phantom{F_{es}} = F_{e\perp}$ for load perpendicular to grain = $6100G^{1.45}/\sqrt{D}$ or NDS Table 8A

$\phantom{F_{es}} = F_{e\theta}$ for load at angle to grain θ (See the Hankinson formula.)

$\phantom{F_{es}}$ For a steel side member, $F_{es} = F_u$.

F_{yb} = bending yield strength of bolt (lbf/in^2)

θ = maximum angle of load to grain; $0 \le \theta \le 90°$ for any member in connection

The dowel bearing strength at an angle of load to grain θ is given by the Hankinson formula.

$$F_{e\theta} = \frac{F_{e\parallel}F_{e\perp}}{F_{e\parallel}\sin^2\theta + F_{e\perp}\cos^2\theta}$$

C. Use of Yield Limit Equations

NDS Eq. 8.2-2 is not used to evaluate wood-to-metal connections. The designer is responsible for making sure that the bearing capacity of the steel side member is not exceeded as required by recognized steel design practices.

As an alternative to these equations, NDS Tables 8.2A through 8.2D contain nominal design values for single-shear connections for common cases. For those cases not in these tables, the previous equations must be used.

The NDS tables may be helpful in providing nominal bolt design values for a number of loading conditions without the need to evaluate the yield limit formulas. However, the NDS tables are restricted to parallel- and perpendicular-to-grain loadings, and the yield limit equations must be evaluated for connections with an intermediate angle of load to grain ($0 < \theta < 90°$).

In the NDS tables,

- For load parallel to grain in both main member and side member, the notation is $Z_\parallel$.
- For load perpendicular to grain in side member and parallel to grain in main member, the notation is $Z_{s\perp}$.
- For load perpendicular to grain in main member and parallel to grain in side member, the notation is $Z_{m\perp}$.

For wood-to-steel connections, the NDS tables provide two values of Z based on the angle of load to grain for the wood member.

- For load parallel to grain in wood main member, the notation is $Z_\parallel$.
- For load perpendicular to grain in wood main member, the notation is $Z_\perp$.

Metal side plates are assumed to be $^1/_4$ in thick ASTM A36 steel.

2. Allowable Design Values for Single-Shear Connections and Adjustment Factors

[NDS Table 7.3.1]

The allowable design values are

$$Z' = \text{(nominal design values)} \\ \times \text{(product of adjustment factors)} \\ = ZC_DC_MC_tC_gC_\Delta$$

C_g = group action factor for connections (See NDS 7.3.6, NDS Eq. 7.3-1, and NDS Table 7.3.6.)

C_Δ = geometry factor for base dimensions of bolts (See NDS 8.5.2 through 8.5.6.)

Nominal design values by the equations or tables are for bolts with edge and end distances and spacings equal to or greater than the minimum required for full design values (see NDS Tables 8.5.4 and 8.5.5). When these edge and end distances and spacings are less than the maximum but greater than the minimum for reduced design values, the nominal design values are multiplied by the smallest geometry factor obtained by the following two equations.

$$C_\Delta = \frac{\text{actual end distance}}{\substack{\text{minimum end distance for} \\ \text{"full" design value}}}$$

$$C_\Delta = \frac{\text{actual bolt spacing}}{\substack{\text{minimum spacing for} \\ \text{"full" design value}}}$$

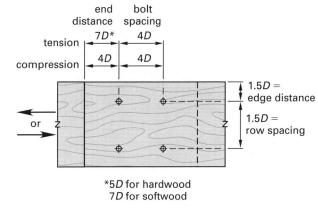

parallel to grain loading

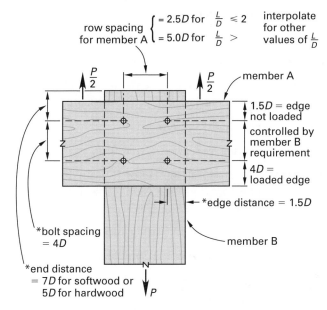

*denotes member B

perpendicular to grain loading (member A)

Figure 9.2 Edge, End, and Spacing Requirements for Bolts (NDS Tables 8.5.3–8.5.6)

Adapted from *National Design Specification for Wood Construction*, 1997 Edition, courtesy of American Forest & Paper Association, Washington, D.C.

Example 9.1(a)
Wood-to-Wood Single-Shear Connection by NDS Tables for Sawn Lumber

A two-member (single-shear) joint is made with 2×6 and 3×6 douglas fir-larch members, select structural grade lumber, and four $^3/_4$ in bolts in $^1/_8$ in oversized holes as shown. The tension load is caused by DL + WL. Assume normal temperature ($C_t = 1.0$) and wet service conditions.

Determine the allowable tension load, P_{allow}, by using NDS table values (not by using the NDS equations).

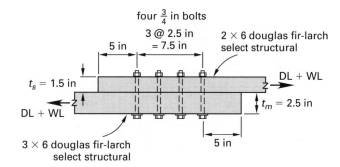

four $\frac{3}{4}$ in bolts

3 @ 2.5 in = 7.5 in

5 in

2 × 6 douglas fir-larch select structural

$t_s = 1.5$ in

DL + WL

DL + WL

$t_m = 2.5$ in

5 in

3 × 6 douglas fir-larch select structural

Solution:

NDS Section

A. Tension load as determined by bolt capacity

The allowable bolt design value is

Table 7.3.1 $Z' = ZC_DC_MC_tC_gC_\Delta$

Table 8.2A For $t_m = 2\frac{1}{2}$ in, $t_s = 1\frac{1}{2}$ in, and $D = \frac{3}{4}$ in, douglas fir-larch: $Z = Z_\parallel = 1020$ lbf

2.3.2 For wind load, $C_D = 1.6$.

Table 7.3.3 For wet in service, $C_M = 0.70$.

Table 7.3.6A The group action factor, C_g, for four bolts in a row (by using NDS Table 7.3.6A, conservatively) is

$$A_m = (2.5 \text{ in})(5.5 \text{ in}) = 13.75 \text{ in}^2$$

$$A_s = (1.5 \text{ in})(5.5 \text{ in}) = 8.25 \text{ in}^2$$

$$\frac{A_s}{A_m} = \frac{8.25 \text{ in}^2}{13.75 \text{ in}^2} = 0.6$$

Table 7.3.6A After interpolation, conservatively using NDS Table 7.3.6A for $\frac{3}{4}$ in bolts (see Table Ftns.), $C_g = 0.88$.

8.5.2 Find the geometry factor, C_Δ.

8.5.4 Find the end distance factor.

The actual end distance is 5.0 in.

Table 8.5.4 For full design value,
$$7D = (7)(0.75 \text{ in})$$
$$= 5.25 \text{ in}$$

For reduced design value,
$$3.5D = (3.5)(0.75 \text{ in})$$
$$= 2.625 \text{ in}$$

$$C_\Delta = \frac{\text{actual end distance}}{\text{minimum end distance required}}$$
$$= \frac{5.0 \text{ in}}{5.25 \text{ in}}$$
$$= 0.952$$

8.5.5 Find the spacing factor. The actual bolt spacing is 2.5 in.

Table 8.5.5 For full design value,
$$4D = (4)(0.75 \text{ in})$$
$$= 3.0 \text{ in}$$

For reduced design value,
$$3D = (3)(0.75 \text{ in})$$
$$= 2.25 \text{ in}$$

$$C_\Delta = \frac{\text{actual spacing}}{\text{minimum spacing required}}$$
$$= \frac{2.5 \text{ in}}{3.0 \text{ in}}$$
$$= 0.833 \quad \text{[controls]}$$

8.5.3 Determine the minimum required edge distance.

$$l_m = t_m = 2.5 \text{ in}$$

$$l_s = t_s = 1.5 \text{ in}$$

$$\frac{l_m}{D} = \frac{2.5 \text{ in}}{0.75 \text{ in}} = 3.33$$

$$\frac{l_s}{D} = \frac{1.5 \text{ in}}{0.75 \text{ in}} = 2.0 \quad \text{[controls for } l/D\text{]}$$

Table 8.5.3 The minimum edge distance required for $l/D = 2.0$ is $1.5D$.

$$1.5D = (1.5)(0.75 \text{ in}) = 1.12 \text{ in}$$

The actual edge distance is
$$\frac{5.5 \text{ in}}{2} = 2.75 \text{ in} > 1.12 \text{ in} \quad \text{[OK]}$$

8.5.6 Row spacing is not applicable.

Table 7.3.1 $Z' = (1020 \text{ lbf})(1.6)(0.70)(1.0)(0.88)(0.833)$
$$= 837.4 \text{ lbf}$$

The allowable tension load as determined by the bolts is

$$P_{\text{allow}} = \left(837.4 \ \frac{\text{lbf}}{\text{bolt}}\right)(4 \text{ bolts}) = 3349.6 \text{ lbf}$$

B. Tension load as determined by wood tension capacity (for 2 × 6 member)

Table 2.3.1 $F_t' = F_tC_DC_MC_tC_FC_i$

NDS
Supp.
Table 4A; $F_t = 1000 \text{ lbf/in}^2$; $C_M = 1.0$
Adj. Factors
$C_F = 1.3$ for 2 × 6 or 3 × 6

$$F_t' = \left(1000 \ \frac{\text{lbf}}{\text{in}^2}\right)(1.6)(1.0)(1.0)(1.3)(1.0)$$
$$= 2080 \ \text{lbf/in}^2$$

$$P_{\text{allow}} = F_t' A_n$$
$$= \left(2080 \ \frac{\text{lbf}}{\text{in}^2}\right)$$
$$\times \big(5.5 \ \text{in} - (0.75 \ \text{in} + 0.125 \ \text{in})\big)$$
$$\times (1.5 \ \text{in})$$
$$= 14{,}430 \ \text{lbf}$$

C. Allowable tension load

$$P_{\text{allow}} = 3349.6 \ \text{lbf}$$

Example 9.1(b)
Wood-to-Wood Single-Shear Connection by NDS Tables for Glulam

Given the same configuration as Ex. 9.1(a) except that the main member is a $2\frac{1}{2}$ in × 6 in glulam tension member made of combination 28 douglas fir-larch, determine the allowable tension load, P_{allow}, by using NDS table values (not by using the NDS equations).

Solution:
NDS Section

A. Bolt capacity

Table 8.2C $\quad t_m = 2\frac{1}{2}$ in (glulam)
$t_s = 1\frac{1}{2}$ in (2 × 6)
$\frac{3}{4}$ in bolts

For douglas fir-larch,
$Z = Z_{\|} = 1020.0 \ \text{lbf}$

$$Z' = Z C_D C_M C_t C_g C_\Delta \quad \begin{bmatrix} \text{refer to Ex. 9.1(a)} \\ \text{for adj. factors} \end{bmatrix}$$
$$= (1020 \ \text{lbf})(1.6)(0.70)(1.0)(0.88)(0.833)$$
$$= 837.4 \ \text{lbf per bolt}$$

$$P_{\text{allow}} = Z'(4 \ \text{bolts})$$
$$= \left(837.4 \ \frac{\text{lbf}}{\text{bolt}}\right)(4 \ \text{bolts})$$
$$= 3349.7 \ \text{lbf}$$

Table 4A ### B. Side member (2 × 6) wood capacity

$$F_t' = F_t C_D C_M C_t C_F C_i$$
$$= 2080 \ \text{lbf/in}^2 \quad \text{[from Ex. 9.1(a)]}$$

$$P_{\text{allow}} = F_t' A_n$$
$$= 14{,}430 \ \text{lbf} \quad \text{[from Ex. 9.1(a)]}$$

C. Allowable tension capacity

$$P_{\text{allow}} = 3349.7 \ \text{lbf}$$

Example 9.2
Wood-to-Wood Single-Shear Connection by NDS Equations for Sawn Lumber

A joint is made as shown.

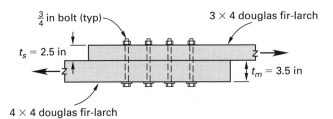

Determine the nominal design value for a single bolt in shear using NDS equations and NDS Tables. Assume all spacing and edge distance requirements are met for full design values.

Solution:
NDS Section

Table 8A $\quad$ For douglas fir-larch, $D = \frac{3}{4}$ in and $G = 0.50$.

$$F_{em\|} = 11{,}200G = (11{,}200)(0.5)$$
$$= 5600 \ \text{lbf/in}^2$$

8.2.1 $\quad \theta = 0°$

$$F_{em} = F_{es} = F_{es\|} = F_{em\|} = 5600 \ \text{lbf/in}^2$$

$$K_\theta = 1 + \frac{\theta}{360°} = 1 + 0 = 1.0$$

$$R_e = \frac{F_{em}}{F_{es}} = 1.0$$

$$R_t = \frac{t_m}{t_s} = \frac{3.5 \ \text{in}}{2.5 \ \text{in}} = 1.4$$

Table 8.2A $\quad F_{yb} = 45{,}000 \ \text{lbf/in}^2$
(Ftn. 2)

Substitution of the above values gives

Eq. 8.2-1 $\quad Z = \dfrac{D t_m F_{em}}{4 K_\theta}$
$$= \frac{\left(\frac{3}{4} \ \text{in}\right)(3.5 \ \text{in})\left(5600 \ \frac{\text{lbf}}{\text{in}^2}\right)}{(4)(1.0)}$$
$$= 3675 \ \text{lbf}$$

Eq. 8.2-2

$$Z = \frac{Dt_sF_{es}}{4K_\theta}$$

$$= \frac{\left(\frac{3}{4}\text{ in}\right)(2.5\text{ in})\left(5600\,\dfrac{\text{lbf}}{\text{in}^2}\right)}{(4)(1.0)}$$

$$= 2625\text{ lbf}$$

$$k_1 = \frac{\sqrt{R_e + 2R_e^2(1 + R_t + R_t^2) + R_t^2R_e^3} - R_e(1 + R_t)}{1 + R_e}$$

$$= \frac{\begin{array}{c}\sqrt{1.0 + (2)(1.0)^2\left(1 + 1.4 + (1.4)^2\right) + (1.4)^2(1.0)^3}\\ - (1.0)(1 + 1.4)\end{array}}{1 + 1.0}$$

$$= 0.509$$

Eq. 8.2-3

$$Z = \frac{k_1Dt_sF_{es}}{3.6K_\theta}$$

$$= \frac{(0.509)\left(\frac{3}{4}\text{ in}\right)(2.5\text{ in})\left(5600\,\dfrac{\text{lbf}}{\text{in}^2}\right)}{(3.6)(1.0)}$$

$$= 1485\text{ lbf}$$

$$k_2 = -1 + \sqrt{(2)(1 + R_e) + \frac{2F_{yb}(1 + 2R_e)D^2}{3F_{em}t_m^2}}$$

$$= -1 + \sqrt{(2)(1 + 1.0) + \frac{(2)\left(45,000\,\dfrac{\text{lbf}}{\text{in}^2}\right) \times \left(1 + (2)(1.0)\right)\left(\frac{3}{4}\text{ in}\right)^2}{(3)\left(5600\,\dfrac{\text{lbf}}{\text{in}^2}\right)(3.5\text{ in})^2}}$$

$$= 1.177$$

Eq. 8.2-4

$$Z = \frac{k_2Dt_mF_{em}}{(3.2)(1 + 2R_e)K_\theta}$$

$$= \frac{(1.177)\left(\frac{3}{4}\text{ in}\right)(3.5\text{ in})\left(5600\,\dfrac{\text{lbf}}{\text{in}^2}\right)}{(3.2)\left(1 + (2)(1.0)\right)(1.0)}$$

$$= 1802\text{ lbf}$$

$$k_3 = -1 + \sqrt{\frac{(2)(1 + R_e)}{R_e} + \frac{2F_{yb}(2 + R_e)D^2}{3F_{em}t_s^2}}$$

$$= -1 + \sqrt{\frac{(2)(1 + 1.0)}{1.0} + \frac{(2)\left(45,000\,\dfrac{\text{lbf}}{\text{in}^2}\right) \times (2 + 1.0)\left(\frac{3}{4}\text{ in}\right)^2}{(3)\left(5600\,\dfrac{\text{lbf}}{\text{in}^2}\right)(2.5\text{ in})^2}}$$

$$= 1.334$$

Eq. 8.2-5

$$Z = \frac{k_3Dt_sF_{em}}{(3.2)(2 + R_e)K_\theta}$$

$$= \frac{(1.334)\left(\frac{3}{4}\text{ in}\right)(2.5\text{ in})\left(5600\,\dfrac{\text{lbf}}{\text{in}^2}\right)}{(3.2)(2 + 1.0)(1.0)}$$

$$= 1459\text{ lbf}$$

Eq. 8.2-6

$$Z = \frac{D^2}{3.2K_\theta}\sqrt{\frac{2F_{em}F_{yb}}{(3)(1 + R_e)}}$$

$$= \frac{\left(\frac{3}{4}\text{ in}\right)^2}{(3.2)(1.0)}\sqrt{\frac{(2)\left(5600\,\dfrac{\text{lbf}}{\text{in}^2}\right)\left(45,000\,\dfrac{\text{lbf}}{\text{in}^2}\right)}{(3)(1 + 1.0)}}$$

$$= 1611\text{ lbf}$$

The smallest value, 1459 lbf, controls the capacity of a ³⁄₄ in bolt in single shear for this example.

NDS Table 8.2A gives

t_m, in	t_s, in	$Z_\parallel$, lbf
3.5	1.5	1200
3.5	3.5	1610

By interpolation for $t_s = 2.5$ in for a ³⁄₄ in bolt,

$$Z_\parallel = \frac{1200\text{ lbf} + 1610\text{ lbf}}{2}$$

$$= 1405\text{ lbf}\left[\begin{array}{l}\text{as compared with the 1459 lbf}\\\text{determined by NDS equations}\end{array}\right]$$

Example 9.3
Wood-to-Wood Single-Shear Connection at an Angle

An assembly consisting of a 4×6 douglas fir-larch at an angle to grain of $60°$ with a 6×8 southern pine horizontal member is joined by a 1 in bolt as shown.

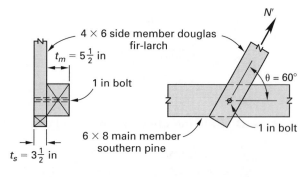

Determine the allowable design value, N', for the joint using NDS tables. Assume that all adjustment factors are 1.0.

Solution:
NDS Section

A. Bolt capacity in the 6 × 8 main southern pine member

Table 8.2A

For $t_m = 5^1/_2$ in and $t_s = 3^1/_2$ in,
$$P = Z_{\parallel} = 2870 \text{ lbf}$$
$$Q = Z_{m\perp} = 1550 \text{ lbf}$$

The Hankinson formula (NDS App. J) gives the nominal lateral design value at an angle to grain as follows.

$$N = \frac{PQ}{P\sin^2\theta + Q\cos^2\theta}$$
$$= \frac{(2870 \text{ lbf})(1550 \text{ lbf})}{(2870 \text{ lbf})(\sin^2 60°) + (1550 \text{ lbf})(\cos^2 60°)}$$
$$= 1751.4 \text{ lbf}$$

Therefore,
$$N' = N(\text{adj. factors}) = (1751.4 \text{ lbf})(1.0)$$
$$= 1751.4 \text{ lbf}$$

B. Bolt capacity in the 4 × 6 side douglas fir-larch member

For douglas fir-larch, $N = Z_{\parallel} = 2660$ lbf.

$Z_{s\perp}$ is not applicable.

Therefore,
$$N' = N(\text{adj. factors}) = (2660 \text{ lbf})(1.0)$$
$$= 2660 \text{ lbf}$$

C. Allowable bolt capacity

The lowest N' controls. Therefore,
$$N' = 1751.4 \text{ lbf}.$$

Example 9.4
Wood-to-Metal Single-Shear Connection at an Angle

A wood-to-metal connection is shown.

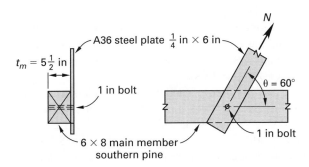

Determine the nominal design value, N, for the connection by using NDS equations and tables.

(Note: The NDS tables can only be used for approximations because of simplifying assumptions for certain cases. However, use of these table values is usually sufficient for ordinary designs.)

Solution:
NDS Section

A. Bolt capacity in the 6 × 8 southern pine main member $(t_m = 5^1/_2 \text{ in})$

By NDS table values,

Table 8.2B

$$P = Z_{\parallel} = 2880 \text{ lbf}$$
$$Q = Z_{\perp} = 1290 \text{ lbf}$$
$$\theta = 60°$$

Use the Hankinson formula.
$$N = \frac{PQ}{P\sin^2\theta + Q\cos^2\theta}$$
$$= \frac{(2880 \text{ lbf})(1290 \text{ lbf})}{(2880 \text{ lbf})(\sin^2 60°) + (1290 \text{ lbf})(\cos^2 60°)}$$
$$= 1496.6 \text{ lbf}$$

By checking all equations, NDS Eq. 8.2-3 controls. (Recall that only NDS Eq. 8.2-2 does not apply to wood-to-metal connections and does not have to be checked.)

Eq. 8.2-3

$$Z = \frac{k_1 D t_s F_{es}}{3.6 K_\theta}$$

Table 8A

$G = 0.55$ for southern pine, and $D = 1$ in.

$$F_{em\parallel} = 6150 \text{ lbf/in}^2$$

$$F_{em\perp} = 2550 \text{ lbf/in}^2$$

$$t_s = 0.25 \text{ in}$$

For ASTM A36 steel, $F_{es} = F_u = 58{,}000 \text{ lbf/in}^2$.

From the Hankinson formula,

$$F_{e\theta} = \frac{F_{em\parallel} F_{em\perp}}{F_{em\parallel}\sin^2\theta + F_{em\perp}\cos^2\theta}$$
$$= \frac{\left(6150 \dfrac{\text{lbf}}{\text{in}^2}\right)\left(2550 \dfrac{\text{lbf}}{\text{in}^2}\right)}{\left(6150 \dfrac{\text{lbf}}{\text{in}^2}\right)(\sin^2 60°) + \left(2550 \dfrac{\text{lbf}}{\text{in}^2}\right)(\cos^2 60°)}$$
$$= 2987.1 \text{ lbf/in}^2$$

$$R_e = \frac{F_{em}}{F_{es}} = \frac{F_{e\theta}}{F_{es}} = \frac{2987.1 \dfrac{\text{lbf}}{\text{in}^2}}{58{,}000 \dfrac{\text{lbf}}{\text{in}^2}} = 0.052$$

$$R_t = \frac{t_m}{t_s} = \frac{5.5 \text{ in}}{0.25 \text{ in}} = 22.0$$

$$K_\theta = 1 + \frac{\theta}{360°} = 1 + \frac{60°}{360°} = 1.167$$

$$k_1 = \frac{\sqrt{R_e + 2R_e^2(1 + R_t + R_t^2) + R_t^2 R_e^3} - R_e(1 + R_t)}{1 + R_e}$$

$$= \frac{\sqrt{\begin{array}{c} 0.052 + (2)(0.052)^2(1 + 22 + (22)^2)) \\ + (22)^2(0.052)^3 \end{array}} - (0.052)(1 + 22)}{1 + 0.052}$$

$$= 0.471$$

$$Z = \frac{k_1 D t_s F_{es}}{3.6 K_\theta}$$

$$= \frac{(0.471)(1 \text{ in})(0.25 \text{ in})\left(58{,}000 \dfrac{\text{lbf}}{\text{in}^2}\right)}{(3.6)(1.167)}$$

$$= 1625.6 \text{ lbf}$$

$$N = Z$$
$$= 1625.6 \text{ lbf} \begin{bmatrix} \text{as compared with 1496.6 lbf} \\ \text{by the NDS table values} \end{bmatrix}$$

(Note: For a thorough analysis, the load capacity of the A36 steel plate must also be checked.)

3. Nominal Design Values for Double-Shear Connections

[NDS 8.3; Eqs. 8.3-1–8.3-4; Tables 8.3A–8.3D]

The yield limit equations for bolt connections in double shear are different from those for bolts in single shear since these two cases have different yield modes (see NDS Fig. 8C and App. I).

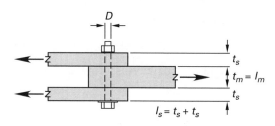

Figure 9.3 Double-Shear Connection

A. Wood-to-Wood Connections

The nominal design value for one bolt in a double-shear connection between three wood members is the smallest load capacity obtained from the following four equations. The thicknesses of the side members are assumed to be equal.

$$Z = \frac{D t_m F_{em}}{4 K_\theta} \qquad \text{[NDS Eq. 8.3-1]}$$

$$Z = \frac{D t_s F_{es}}{2 K_\theta} \qquad \text{[NDS Eq. 8.3-2]}$$

$$Z = \frac{k_3 D t_s F_{em}}{(1.6)(2 + R_e) K_\theta} \qquad \text{[NDS Eq. 8.3-3]}$$

$$Z = \frac{D^2}{1.6 K_\theta} \sqrt{\frac{2 F_{em} F_{yb}}{(3)(1 + R_e)}} \qquad \text{[NDS Eq. 8.3-4]}$$

$$k_3 = -1 + \sqrt{\frac{(2)(1 + R_e)}{R_e} + \frac{2 F_{yb}(2 + R_e) D^2}{3 F_{em} t_s^2}}$$

$$R_e = \frac{F_{em}}{F_{es}}$$

$$K_\theta = 1 + \frac{\theta}{360°}$$

F_{em} = dowel bearing strength of main (center) member (lbf/in^2) (See NDS Table 8A.)
  = $F_{e\parallel}$ for load parallel to grain
  = $F_{e\perp}$ for load perpendicular to grain
  = $F_{e\theta}$ for load at angle to grain θ
  (See the Hankinson formula.)

F_{es} (See NDS Table 8A.)
  = $F_{e\parallel}$ for load parallel to grain
  = $F_{e\perp}$ for load perpendicular to grain
  = $F_{e\theta}$ for load at angle to grain θ (see Hankinson formula). For a steel side member, $F_{es} = F_u$.

F_{yb} = bending yield strength of bolt (lbf/in^2)

θ = maximum angle of load to grain ($0 \leq \theta \leq 90°$) for any member in connection

The dowel bearing strength at an angle of load to grain θ is given by the Hankinson formula.

$$F_{e\theta} = \frac{F_{e\parallel} F_{e\perp}}{F_{e\parallel} \sin^2 \theta + F_{e\perp} \cos^2 \theta}$$

B. Wood-to-Metal Connections

For one bolt in a double-shear connection between a wood main member and two steel side plates, the nominal design value is taken as the smallest load capacity obtained from the three equations previously given for a wood-to-wood connection (NDS Eqs. 8.3-1, 8.3-3, and 8.3-4). NDS Eq. 8.3-2 is not used to evaluate this type of wood-to-metal connection. The designer is responsible for making sure that the bearing capacity of the steel side members is not exceeded as required by prevailing steel design practices.

A less frequently encountered double-shear connection involves a steel main (center) member and two wood side members. In this situation, NDS Eqs. 8.3-2, 8.3-3, and 8.3-4 are applicable. NDS Eq. 8.3-1 is not used to evaluate this type of wood-to-metal connection. The designer is responsible for making sure that the bearing capacity of the steel main member is not exceeded as required by prevailing steel design practices.

C. NDS Tables

As an alternative to the previous four equations, NDS Tables 8.3A through 8.3D provide nominal design values for double-shear connections for most common cases. The cases not included in these tables must be determined by the equations.

4. Nominal Design Values for Multiple-Shear Connections
[NDS 8.4 and 8.2]

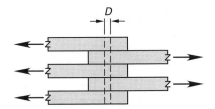

Figure 9.4 Multiple-Shear Connection

The nominal design value for a multiple-shear connection is the lowest value obtained for a single-shear plane multiplied by the number of shear planes.

Example 9.5(a)
Wood-to-Wood Double-Shear Connection for Sawn Lumber

A three-member joint is made of douglas fir-larch, select structural grade lumbers, 2×6 for side members and 3×6 for the main member, with two $^3/_4$ in bolts as shown. A wet service condition exists. Assume all other adjustment factors are 1.0.

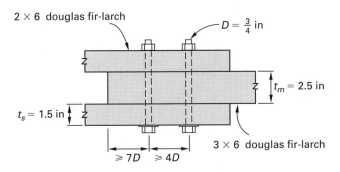

Determine the allowable tension load.

Solution:

NDS Section

A. Bolt capacity

Table 7.3.1 $Z' = ZC_DC_MC_tC_gC_\Delta$

Table 7.3.3 $C_M = 0.70$

Table 8.3A $Z = 2400$ lbf

For $t_m = 2^1/_2$ in, $t_s = 1^1/_2$ in, $^3/_4$ in bolt, and douglas fir-larch,

$$Z' = (2400 \text{ lbf})(1.0)(0.70)(1.0)(1.0)(1.0)$$
$$= 1680 \text{ lbf/bolt}$$

$$P_{\text{allow}} = Z'(\text{no. of bolts})$$
$$= \left(1680 \ \frac{\text{lbf}}{\text{bolt}}\right)(2 \text{ bolts})$$
$$= 3360 \text{ lbf}$$

B. Lumber capacity
(3×6 controls the capacity)

Table 2.3.1 $F_t' = F_tC_DC_MC_tC_F$

Supp. $F_t = 1000 \text{ lbf/in}^2$
Table 4A
               $C_M = 1.0$

               $C_F = 1.3$

$$F_t' = \left(1000 \ \frac{\text{lbf}}{\text{in}^2}\right)(1.0)(1.0)(1.0)(1.3)$$
$$= 1300 \text{ lbf/in}^2$$

$$P_{\text{allow}} = F_t'A_n$$
$$= \left(1300 \ \frac{\text{lbf}}{\text{in}^2}\right)(2.5 \text{ in})$$
$$\times \left(5.5 \text{ in} - (0.75 \text{ in} + 0.125 \text{ in})\right)$$
$$= 15{,}031 \text{ lbf}$$

C. The allowable tension load is the lowest P_{allow}

$$P_{\text{allow}} = 3360 \text{ lbf}$$

Example 9.5(b)
Wood-to-Metal Double-Shear Connection for Glulam

A three-member tension joint consists of a $2^1/_2$ in $\times$ 6 in glulam main member, combination 28 douglas fir, and two 6 in wide by $^1/_4$ in thick A36 steel side plates.

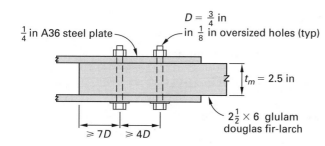

Determine the allowable tension load as based on bolt and glulam member capacities.

Solution:

NDS Section

A. Bolt capacity

Table 7.3.1 $Z' = ZC_DC_MC_tC_gC_\Delta$ [from Ex. 9.5(a)]

Table 7.3.3 $C_M = 0.70$

Table 8.3D   For $t_m = 2\frac{1}{2}$ in, $t_s = \frac{1}{4}$ in, douglas fir glulam combination, and $\frac{3}{4}$ in bolt,

$$Z = Z_\parallel = 2630 \text{ lbf}$$

$$Z' = (2630 \text{ lbf})(1.0)(0.70)(1.0)(1.0)(1.0)$$
$$= 1841 \text{ lbf/bolt}$$

$$P_{\text{allow}} = Z'(2 \text{ bolts})$$
$$= \left(1841 \, \frac{\text{lbf}}{\text{bolt}}\right)(2 \text{ bolts})$$
$$= 3682 \text{ lbf}$$

B. For glulam, 28 douglas fir, $2\frac{1}{2}$ in × 6 in capacity

$2\frac{1}{2}$ in × 6 in is made up of four laminations.

Supp. Table 1C
Supp. Table 5B

$$F_t = 1100 \text{ lbf/in}^2$$

$$C_M = 0.8$$

Table 2.3.1 $F_t' = F_tC_DC_MC_t$ for glulam members

$$= \left(1100 \, \frac{\text{lbf}}{\text{in}^2}\right)(1.0)(0.8)(1.0)$$
$$= 880 \text{ lbf/in}^2$$

$$P_{\text{allow}} = F_t'A_n$$
$$= \left(800 \, \frac{\text{lbf}}{\text{in}^2}\right)(2.5 \text{ in})$$
$$\times \left(6 \text{ in} - (0.75 \text{ in} + 0.125 \text{ in})\right)$$
$$= 11{,}275 \text{ lbf}$$

C. Allowable tension load = smallest P_{allow}

$$P_{\text{allow}} = 3682 \text{ lbf}$$

(Note: Although this problem did not require that the tension capacities of the steel side members be checked, they should be checked in an actual design and/or analysis situation.)

Lag Screws and Wood Screws

1. Lag Screws (Lag Bolts)

[NDS Part 9]

Lag screws, also called *lag bolts*, are more similar to bolts than they are to wood screws. Their sizes range from $1/4$ in to $1^1/4$ in in diameter with lengths varying from 1 in to 12 in (NDS Part 9, NDS App. L). Wood screw diameters range from $1/8$ in to $3/8$ in (circular round or flat head and $1/2$ in to 4 in long).

Lag screws, which have a hex or square head, consist of nonsmooth and smooth shanks. Their abilities to resist either withdrawal or lateral loads are superior to those of nails.

Lag screws are inserted into lead holes. However, lead holes are not required for $3/8$ in and smaller diameter lag screws loaded primarily in withdrawal (NDS 9.1.2).

2. Withdrawal Design Values for Lag Screws

[NDS 9.2; Tables 7.3.1, 9A, and 9.2A]

The nominal design value for withdrawal specified in NDS Table 9.2A is the withdrawal value per inch of thread penetration into the side grain of the main (holding) member.

The allowable withdrawal design value is

$$W' = \text{(nominal design values)} \\ \times \text{(product of adjustment factors)} \\ = WpC_D C_M C_t C_{eg}$$

W = nominal withdrawal design value, lbf/in of thread penetration into side grain (See NDS Table 9.2A.)

p = effective threaded penetration (in)

C_D (See NDS 2.3.2, 7.3.2.)

C_M (See NDS Table 7.3.3.)

C_t (See NDS 7.3.4.)

C_{eg} = end grain factor if loaded in withdrawal from end grain
= 0.75 (See NDS 9.2.2.)

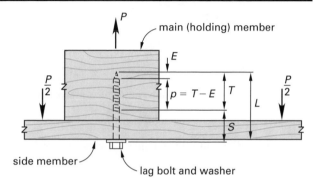

Figure 10.1 Lag Screws: Effective Thread Penetration into Main Member, $p = T - E$ (based on NDS App. L)

Adapted from *National Design Specification for Wood Construction*, 1997 Edition, courtesy of American Forest & Paper Association, Washington, D.C.

Example 10.1
Lag Bolt Connection for Withdrawal

A $1/2$ in × 6 in lag bolt is used to attach a 4 × 16 side member to a 12 × 12 beam as shown. Both members are made of southern pine. The lumber is exposed to weather in service. Assume that the load is permanent ($C_D = 0.9$ as per NDS 2.3.2) and the connection is under normal temperature conditions ($C_t = 1.0$).

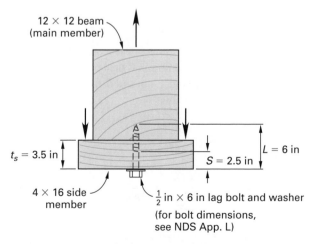

Determine the allowable withdrawal load.

Solution:

NDS Section

App. L For a $\frac{1}{2}$ in $\times$ 6 in lag bolt with an $\frac{1}{8}$ in washer, the length of lag bolt into beam (main member) is

$$6 \text{ in} - 3.5 \text{ in (for side member)}$$
$$- 0.125 \text{ in (for washer)}$$
$$= 2.375 \text{ in}$$

App. L The length of the tapered tip is $E = \frac{5}{16}$ in.

The effective thread penetration length into beam (main member) is

$$p = 2.375 \text{ in} - E$$
$$= 2.375 \text{ in} - \frac{5}{16} \text{ in}$$
$$= 2.06 \text{ in}$$

Table 7.3.3 $C_M = 0.70$ [exposed to weather in-service]

9.2.2 $C_{eg} = 1.0$ [no end grain application]

Table 9A $G = 0.55$ for southern pine

Table 9.2A $W = 437$ lbf per inch effective penetration

The allowable design load is

$$W' = WpC_DC_MC_tC_{eg}$$
$$= \left(437 \, \frac{\text{lbf}}{\text{in}}\right)(2.06 \text{ in})(0.9)(0.70)(1.0)(1.0)$$
$$= 567.1 \text{ lbf}$$

3. Lateral Design Values

[NDS 9.3; Tables 9.3A and 9.3B]

The yield limit equations for lag bolts in single shear are similar to those for bolt connections. However, there are fewer equations; only NDS Eqs. 9.3-1 through 9.3-3 are necessary.

A. Wood-to-Wood Connection

For one lag bolt in a single-shear connection between two wood members, the nominal design value is taken as the smallest load capacity obtained from the following three yield limit equations.

$$Z = \frac{Dt_sF_{es}}{4K_\theta} \qquad \text{[NDS Eq. 9.3-1]}$$

$$Z = \frac{kDt_sF_{em}}{(2.8)(2 + R_e)K_\theta} \qquad \text{[NDS Eq. 9.3-2]}$$

$$Z = \frac{D^2}{3K_\theta}\sqrt{\frac{1.75F_{em}F_{yb}}{(3)(1 + R_e)}} \qquad \text{[NDS Eq. 9.3-3]}$$

$$k = -1 + \sqrt{\frac{(2)(1 + R_e)}{R_e} + \frac{F_{yb}(2 + R_e)D^2}{2F_{em}t_s^2}}$$

$$R_e = \frac{F_{em}}{F_{es}}$$

$$K_\theta = 1 + \frac{\theta}{360°}$$

$D =$ diameter of unthreaded shank of lag bolt (in)

F_{em} (See NDS Table 9A.)
$= F_{e\parallel}$ for parallel-to-grain loading
$= F_{e\perp}$ for perpendicular-to-grain loading
$= F_{e\theta}$ for load at angle to grain θ
(See the Hankinson formula.)

F_{es} (See NDS Table 9A.)
$= F_{e\parallel}$ for parallel-to-grain loading
$= F_{e\perp}$ for perpendicular-to-grain loading
$= F_{e\theta}$ for load at angle to grain θ
(See the Hankinson formula.)
For a steel side member, $F_{es} = F_u$.

F_{yb} (See Ftns. of NDS Table 9.3A.)
$= 70,000$ lbf/in^2 [for $D = \frac{1}{4}$ in]
$= 60,000$ lbf/in^2 [for $D = \frac{5}{16}$ in]
$= 45,000$ lbf/in^2 [for $D \geq \frac{3}{8}$ in]

$\theta =$ maximum angle of load to grain ($0 \leq \theta \leq 90°$) for any member in connection.

$$F_{e\theta} = \frac{F_{e\parallel}F_{e\perp}}{F_{e\parallel}\sin^2\theta + F_{e\perp}\cos^2\theta}$$
(See the Hankinson formula.)

B. Wood-to-Metal Connection

For one lag bolt in a single-shear connection between a wood main member and a steel side plate, the nominal design value is taken as the smallest load capacity obtained from two of the three yield limit equations previously given. NDS Eq. 9.3-1 is eliminated for wood-to-metal connections. In applying the yield limit equations, F_{es} may conservatively be taken as equal to the ultimate tensile strength, F_u, of the steel side plate. The designer is responsible for ensuring that the bearing capacity of the steel side member is not exceeded in accordance with recognized steel design practices.

As an alternative to solving these yield limit equations, the nominal design values for lag bolts used in common applications are given in NDS Tables 9.3A and 9.3B. These values are given for parallel- and perpendicular-to-grain loadings with the following definitions of the notations used.

$Z_\parallel$ = loading parallel to grain in both main and side members

$Z_{s\perp}$ = loading perpendicular to grain in side member and parallel to grain in main member

$Z_{m\perp}$ = loading perpendicular to grain in main member and parallel to grain in side member

The nominal design values in the NDS tables are for the base conditions: dry at fabrication and in service, normal temperature, penetration of lag bolt into the main member with a minimum penetration length of eight times the size of the shank diameter (see NDS 9.3.3), and other parameters necessary for base conditions.

C. Allowable Design Value, Z'

The allowable design values are given by

$$Z' = \begin{array}{l}\text{(nominal design values)} \\ \times \text{(product of adjustment factors)}\end{array}$$
$$= Z C_D C_M C_t C_g C_\Delta C_d C_{eg}$$
(See NDS Table 7.3.1.)

C_D, C_M, C_t (See NDS 2.3.2 to 2.3.4 and 7.3.)

C_g (See NDS 7.3.6, Tables 7.3.6A and 7.3.6C.)

C_Δ = geometry factor for bolt base dimensions (See NDS 8.5.2 through 8.5.6.)

C_d (See NDS 9.3.3.)

$C_{eg} = 0.67$ (See NDS 9.3.4.)

Geometry Factor, C_Δ
[NDS 8.5.2 through 8.5.6]

End distances, edge distances, and spacing requirements for lag bolts are the same as for bolted connections.

Penetration Depth Factor, C_d
[NDS 9.3.3; Eq. 9.3-5; App. L]

For penetration depth factor, use

$$C_d = \frac{p}{8D} \le 1.0$$

p = penetration depth when $4D \le p \le 8D$
$\quad = L - t_s - t_{\text{washer}} - E$

L = nominal length

t_s = thickness of side (thinner) member

E = length of tapered tip

D = shank diameter (of unthreaded length)

Example 10.2
Lateral Loads on Lag Screws for Wood-to-Metal Connection

In the figure shown, two $^5/_8$ in × 5 in lag screws are used to attach a $^1/_4$ in × 4 in steel strap to a $5^1/_8$ in × 6 in douglas fir-larch glulam beam. The beam is exposed to weather. Assume normal temperature ($C_t = 1.0$). The total load consists of dead load and wind load (DL + WL).

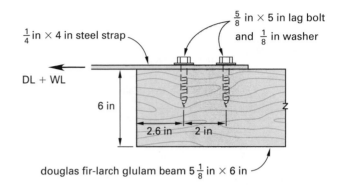

douglas fir-larch glulam beam $5\frac{1}{8}$ in × 6 in

Determine the allowable lateral load that can be carried by the two lag screws.

Solution:

NDS Section The allowable design value is

Table 7.3.1 $Z' = Z C_D C_M C_t C_g C_\Delta C_d C_{eg}$

A. Nominal design value, Z

$\theta = 0°$

Table 9.3A $F_{yb} = 45{,}000 \text{ lbf/in}^2$

Table 9A The specific gravity for the main member is $G = 0.50$.

$F_{em} = F_\parallel = 5600 \text{ lbf/in}^2$

$F_{es} = F_u = 58{,}000 \text{ lbf/in}^2$ for A36 steel

$t_s = 0.25 \text{ in}$

$D = 0.625 \text{ in}$

Eqs. 9.3-1, 9.3-2, 9.3-3
$$R_e = \frac{F_{em}}{F_{es}} = \frac{5600 \, \dfrac{\text{lbf}}{\text{in}^2}}{58{,}000 \, \dfrac{\text{lbf}}{\text{in}^2}} = 0.097$$

$$K_\theta = 1 + \frac{\theta}{360°} = 1 + \frac{0°}{360°} = 1.0$$

$$k = -1 + \sqrt{\frac{(2)(1+R_e)}{R_e} + \frac{F_{yb}(2+R_e)D^2}{2F_{em}t_s^2}}$$

$$= -1 + \left[\frac{(2)(1+0.097)}{0.097} + \frac{\left(45{,}000 \, \dfrac{\text{lbf}}{\text{in}^2}\right)(2+0.097)(0.625 \text{ in})^2}{(2)\left(5600 \, \dfrac{\text{lbf}}{\text{in}^2}\right)(0.25 \text{ in})^2} \right]^{1/2}$$

$$= 7.68$$

Eq. 9.3-1

$$Z = \frac{Dt_s F_{es}}{4K_\theta}$$

$$= \frac{(0.625 \text{ in})(0.25 \text{ in})\left(58{,}000 \ \frac{\text{lbf}}{\text{in}^2}\right)}{(4)(1.0)}$$

$$= 2265.6 \text{ lbf}$$

Eq. 9.3-2

$$Z = \frac{kDt_s F_{em}}{(2.8)(2 + R_e)K_\theta}$$

$$= \frac{(7.68)(0.625 \text{ in})(0.25 \text{ in})\left(5600 \ \frac{\text{lbf}}{\text{in}^2}\right)}{(2.8)(2 + 0.097)(1.0)}$$

$$= 1144.5 \text{ lbf} \quad \text{[controls]}$$

Eq. 9.3-3

$$Z = \frac{D^2}{3K_\theta}\sqrt{\frac{1.75 F_{em} F_{yb}}{(3)(1 + R_e)}}$$

$$= \frac{(0.625 \text{ in})^2}{(3)(1.0)}\sqrt{\frac{(1.75)\left(5600 \ \frac{\text{lbf}}{\text{in}^2}\right)\left(45{,}000 \ \frac{\text{lbf}}{\text{in}^2}\right)}{(3)(1 + 0.097)}}$$

$$= 1507.3 \text{ lbf}$$

Therefore, $Z = 1144.5$ lbf (as compared with 1140.0 lbf by using NDS Table 9.3B for $t_s = 1/4$ in, $D = 5/8$ in, douglas fir-larch).

B. Adjustment factors

2.3.2   For wind loads, $C_D = 1.6$.

Table 7.3.3   For exposed-to-weather condition, $C_M = 0.70$.

Table 7.3.6C Determine the group factor, C_g.

$s = 2.0$ in

For a $5\frac{1}{8}$ in $\times$ 6 in glulam beam,

$A_m = (5.125 \text{ in})(6 \text{ in}) = 30.75 \text{ in}^2$

For a $1/4$ in $\times$ 4 in steel strap,

$A_s = (0.25 \text{ in})(4 \text{ in}) = 1.0 \text{ in}^2$

$$\frac{A_m}{A_s} = \frac{30.75 \text{ in}^2}{1.0 \text{ in}^2} = 30.75$$

For two lag screws in a row, $C_g = 1.0$.

8.5.2   The geometry factor, C_Δ, is as follows.

8.5.3   Check the edge distance.

The minimum edge distance required is

$1.5D = (1.5)(0.625 \text{ in}) = 0.94 \text{ in}$

The actual edge distance is

$$\left(6 \text{ in} - \left(\tfrac{5}{8} \text{ in} + \tfrac{1}{8} \text{ in}\right)\right)\left(\tfrac{1}{2}\right)$$

$$= 2.62 \text{ in} > 1.5D \quad \text{[OK]}$$

8.5.4   Determine the end distance factor.

The minimum end distance required for full design value is

$7D = (7)(0.625 \text{ in}) = 4.38 \text{ in}$

The minimum end distance required for reduced design value is

$3.5D = (3.5)(0.625 \text{ in})$
$= 2.19 \text{ in} < \text{actual end distance}$
$= 2.6 \text{ in} \quad \text{[OK]}$

$$C_\Delta = \frac{\text{actual end distance}}{\text{minimun end distance required}}$$

$$= \frac{2.62 \text{ in}}{4.38 \text{ in}} = 0.598 \quad \text{[controls]}$$

8.5.5   Find the spacing factor, C_Δ.

The minimum spacing required for full design value is

$4D = (4)(0.625 \text{ in}) = 2.5 \text{ in}$

For reduced design value,

$3D = (3)(0.625 \text{ in})$
$= 1.88 \text{ in} < \text{actual spacing}$
$= 2.0 \text{ in} \quad \text{[OK]}$

$$C_\Delta = \frac{\text{actual spacing}}{\text{minimum spacing required}} = \frac{2.0 \text{ in}}{2.5 \text{ in}}$$

$$= 0.80 \quad \text{[controls]}$$

The smaller C_Δ controls. Therefore, use $C_\Delta = 0.598$.

Eq. 9.3-5   Find the penetration depth factor, C_d.

The actual penetration, p, into the beam is

9.3.3; $p = L - t_s - t_{\text{washer}} - E$
App. L $= 5 \text{ in} - \tfrac{1}{4} \text{ in} - \tfrac{1}{8} \text{ in} - \tfrac{13}{32} \text{ in}$

$= 4.22 \text{ in}$

The minimum penetration required for full design value is

$8D = (8)(0.625 \text{ in}) = 5.0 \text{ in}$

For reduced design value,

$4D = (4)(0.625 \text{ in}) = 2.5 \text{ in}$
$2.5 \text{ in} < 4.22 \text{ in}$
$4D < p \quad \text{[OK]}$

$$C_d = \frac{p}{8D} = \frac{4.22 \text{ in}}{5.0 \text{ in}} = 0.844$$

9.3.4   Find the end grain factor, C_{eg}.

Lag screws are inserted into the side grain of the main member. Therefore, $C_{eg} = 1.0$.

Table 7.3.1 **C. Allowable design value, Z'**

$$Z' = ZC_DC_MC_tC_gC_\Delta C_dC_{eg}$$
$$= (1144.5 \text{ lbf})(1.6)(0.70)(1.0)(1.0)$$
$$\times (0.598)(0.844)(1.0)$$
$$= 647 \text{ lbf per lag screw}$$

The allowable lateral load is

$$P_{\text{allow}} = \left(647 \text{ } \frac{\text{lbf}}{\text{screw}}\right)(2 \text{ lag screws})$$
$$= 1294 \text{ lbf}$$

(Note: A thorough analysis requires that the load capacity of the $1/4$ in $\times$ 4 in steel strap be checked, also.)

Example 10.3
Lateral Loads on Lag Screws for Wood-to-Wood Connection

A $1/2$ in $\times$ 6 in lag bolt (screw) connects a $2 \times 6 \times 14$ in long shelf support to a 2×4 stud wall as shown. The total load of 250 lbf consists of dead load and live load (DL + LL). Assume a dry condition and douglas fir-larch lumber.

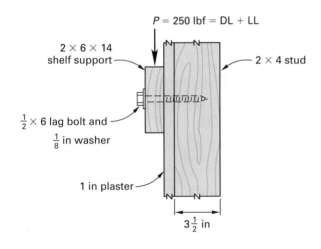

$P = 250$ lbf $=$ DL $+$ LL

$2 \times 6 \times 14$ shelf support

2×4 stud

$\frac{1}{2} \times 6$ lag bolt and $\frac{1}{8}$ in washer

1 in plaster

$3\frac{1}{2}$ in

Determine whether the connection is strong enough to support the total load.

Solution:
NDS Section

The allowable design value is

Table 7.3.1 $Z' = ZC_DC_MC_tC_gC_\Delta C_dC_{eg}$

A. Nominal design value, Z

Table 9.3A $Z = Z_{s\perp} = 380.0$ lbf

B. Adjustment factors

$$C_D = C_t$$
$$= 1.0 \text{ for the given conditions}$$

Table 7.3.3 $C_M = 1.0$

For main member (loaded parallel to grain):

8.5.2 The geometry factor, C_Δ, is found as follows.

8.5.4 C_Δ is determined from the end distance requirement.

The actual end distance provided is several inches.

The minimum end distance required for full design value is

$$4D = (4) \left(\tfrac{1}{2} \text{ in}\right) = 2.0 \text{ in}$$

For reduced design value,

$$2D = (2) \left(\tfrac{1}{2} \text{ in}\right) = 1.0 \text{ in}$$

$$C_\Delta = \frac{\text{end distance}}{2.0 \text{ in}} > 1.0$$

Therefore, $C_\Delta = 1.0$.

8.5.5 The bolt spacing factor, C_Δ, is not applicable.

Therefore, $C_\Delta = 1.0$.

8.5.3 Check the edge distance.

$$\text{actual edge distance} = \frac{1.5 \text{ in}}{2} = 0.75 \text{ in}$$

$$\begin{aligned}\text{required edge distance} &= 1.5D \\ &= (1.5)\left(\tfrac{1}{2} \text{ in}\right) \\ &= 0.75 \text{ in}\end{aligned}$$

Therefore, the edge distance is acceptable.

For side member (loaded perpendicular to grain):

8.5.3 Check the edge distance.

The required loaded edge distance is

$$4D = (4) \left(\tfrac{1}{2} \text{ in}\right) = 2.0 \text{ in}$$

The actual loaded edge distance is

$$\frac{5.5 \text{ in}}{2} = 2.75 \text{ in} > 4D$$

Therefore, the loaded edge distance is acceptable.

The required unloaded edge distance is

$$1.5D = (1.5)\left(\tfrac{1}{2} \text{ in}\right) = 0.75 \text{ in}$$

The actual unloaded edge distance is

$$\frac{5.5 \text{ in}}{2} = 2.75 \text{ in} > 1.5D$$

Therefore, the unloaded edge distance is acceptable.

Table 7.3.6A Find the group action factor, C_g.

Since there is one lag bolt, $C_g = 1.0$.

9.3.3;   Find the penetration depth factor, C_d.

App. L

For $\frac{1}{2}$ in $\times$ 6 in lag bolt, $L = 6$ in; $E = \frac{5}{16}$ in.

The actual penetration into the beam is
$$p = L - t_s - t_{\text{washer}} - 1 \text{ in plaster} - E$$
$$= 6 \text{ in} - \left(1.5 \text{ in} + \tfrac{1}{8} \text{ in}\right) - 1 \text{ in} - \tfrac{5}{16} \text{ in}$$
$$= 3.06 \text{ in}$$

The minimum penetration required for full design value is
$$8D = (8)\left(\tfrac{1}{2} \text{ in}\right) = 4.0 \text{ in}$$

For reduced design value,
$$4D = (4)\left(\tfrac{1}{2} \text{ in}\right) = 2.0 \text{ in}$$

$$2.0 \text{ in} < 3.06 \text{ in}$$
$$4D < p \quad [\text{OK}]$$

Eq. 9.3-5 $C_d = \dfrac{p}{8D} = \dfrac{3.06 \text{ in}}{4.0 \text{ in}} = 0.765$

C. Allowable lag bolt capacity, Z'

Table 7.3.1 $Z' = ZC_D C_M C_t C_g C_\Delta C_d C_{eg}$
$$= (380 \text{ lbf})(1.0)(1.0)(1.0)(1.0)$$
$$\times (0.765)(1.0)$$
$$= 290.7 \text{ lbf} > \text{total load of } 250 \text{ lbf} \quad [\text{OK}]$$

Example 10.4
Lag Bolts Loaded Laterally at an Angle to Grain

A $\frac{1}{2}$ in $\times$ 4 in lag bolt connects a $\frac{1}{4}$ in thick steel plate to a $6\frac{3}{4}$ in $\times$ 11 in southern pine glulam beam (22F-V1 SP/SP) as shown. Assume all adjustment factors, except C_d, are 1.0. Use NDS table values for an approximation.

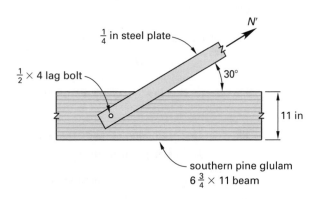

Determine the maximum allowable tensile design force, N', that can be applied to the steel plate as based on bolt and glulam member capacities.

Solution:

NDS Section

Table 7.3.1 $Z' = ZC_D C_M C_t C_g C_\Delta C_d C_{eg}$

Table 9.3B $Z_\parallel = 810.0 \text{ lbf}; Z_\perp = 520.0 \text{ lbf}$

Application of the Hankinson formula gives
$$N = Z = \frac{Z_\parallel Z_\perp}{Z_\parallel \sin^2 \theta + Z_\perp \cos^2 \theta}$$
$$= \frac{(810 \text{ lbf})(520 \text{ lbf})}{(810 \text{ lbf})(\sin^2 30°) + (520 \text{ lbf})(\cos^2 30°)}$$
$$= 710.9 \text{ lbf}$$

Find the penetration depth factor, C_d.

Actual penetration, p, into the beam is

App. L $p = L - t_s - t_{\text{washer}} - E$
$$= 4.0 \text{ in} - 0.25 \text{ in} - 0.0 \text{ in} - \tfrac{5}{16} \text{ in}$$
$$= 3.44 \text{ in}$$

9.3.3   The minimum penetration required for full design value is
$$8D = (8)\left(\tfrac{1}{2} \text{ in}\right) = 4.0 \text{ in}$$

The minimum penetration required for reduced design value is
$$4D = (4)\left(\tfrac{1}{2} \text{ in}\right) = 2.0 \text{ in}$$

$$2.0 \text{ in} < 3.44 \text{ in}$$
$$4D < p \quad [\text{OK}]$$

Eq. 9.3-5 $C_d = \dfrac{p}{8D} = \dfrac{3.44 \text{ in}}{4.0 \text{ in}} = 0.86$

Table 7.3.1   Find the allowable load, N'.

Note that $N = Z$ and $N' = Z'$.

$$N' = N(1.0)(1.0)(1.0)(1.0)(1.0)(0.86)(1.0)$$
$$= (710.9 \text{ lbf})(0.86) = 611.4 \text{ lbf}$$

(Note: Although not required for this problem, the steel plate capacity should also be checked in an actual design and/or analysis situation.)

4. Wood Screws

[NDS Section 11]

Wood screws are similar to lag bolts but are smaller in diameter. Wood screws have either a circular flat head or a circular round head, while lag bolts have either a hex head or a square head as described previously. The load capacities of wood screws are relatively small compared with those of lag bolts.

Wood screws are designated by gage, which is an indicator of the diameter of the shank (the shank is the smooth, unthreaded part of the screw length). For example, no. 10 and no. 20 gages have shank diameters of 0.190 in and 0.320 in, respectively. Approximate sizes for common screws are $^1/_8$ in to $^3/_8$ in in diameter and $^1/_2$ in to 4 in in length.

5. Withdrawal Design Values for Wood Screws

[NDS 11.2; NDS Tables 7.3.1, 11A, and 11.2A]

The nominal withdrawal design values for a single wood screw are listed in NDS Table 11.2A and are in units of pounds per inch of threaded penetration into the side grain of the main member (threaded length is approximately two-thirds of the total screw length). NDS tabulated nominal design values, W, are multiplied by all applicable adjustment factors (NDS Table 7.3.1) to obtain the allowable design values, W'. Wood screws are not allowed to be loaded in withdrawal from end grain of wood.

$W' = W_p$ (product of adjustment factors)

$\quad = WpC_DC_MC_t$

$W \quad$ (See NDS Tables 11A and 11.2A.)

$C_D, C_t \quad$ (See NDS 2.3.)

$\quad C_M \quad$ (See NDS Table 7.3.3.)

$\quad p = $ effective thread penetration (in)

Example 10.5
Withdrawal Load for Wood Screws

Two no. 12 gage (0.216 in diameter, see NDS Table 11.3A) $\times$ 3 in long wood screws connect a $^3/_8$ in thick steel plate to a 12×12 southern pine beam as shown. Assume a wet service condition and normal temperature ($C_t = 1.0$).

Determine the allowable load for a one-year duration. Assume spacing requirements and edge distance requirements are fully satisfied.

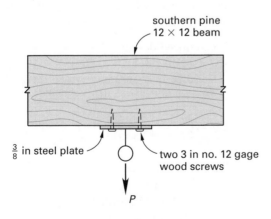

southern pine
12×12 beam

$^3/_8$ in steel plate

two 3 in no. 12 gage
wood screws

P

Solution:

NDS Section

Table 11A   For southern pine, $G = 0.55$.

Table 11.2A $W = 186$ lbf per inch of threaded penetration length

The penetration into the beam is

3.0 in $- \frac{3}{8}$ in steel $= 2.625$ in

As per NDS Table 11.2A subtitle, the *threaded* penetration length is approximately equal to

$\left(\dfrac{2}{3}\right)(3.0 \text{ in}) = 2.0 \text{ in} < 2.625 \text{ in}$

Use $p = 2.0$ in.

Table 7.3.1 $W' = WpC_DC_MC_t$

2.3.2; App. B $C_D = 1.1$

Table 7.3.3 $C_M = 0.70$

$$W' = \left(186 \ \frac{\text{lbf}}{\text{in}}\right)(2 \text{ in})(1.1)(0.70)(1.0)$$

$$= 286.4 \text{ lbf}$$

The allowable load is

$$P_{\text{allow}} = W'(2 \text{ screws})$$

$$= \left(286.4 \ \frac{\text{lbf}}{\text{screw}}\right)(2 \text{ screws})$$

$$= 572.8 \text{ lbf}$$

6. Lateral Design Values for Wood Screws

[NDS 11.3; NDS Tables 11A, 11.3A, and 11.3B]

The yield limit equation for the nominal wood screw lateral design value, Z, will be the lesser of NDS Eqs. 11.3-1, 11.3-2, and 11.3-3.

$$Z = \frac{Dt_sF_{es}}{K_D} \qquad \text{[NDS Eq. 11.3-1]}$$

$$Z = \frac{kDt_sF_{em}}{K_D(2 + R_e)} \qquad \text{[NDS Eq. 11.3-2]}$$

$$Z = \frac{D^2}{K_D}\sqrt{\frac{1.75F_{em}F_{yb}}{(3)(1 + R_e)}} \qquad \text{[NDS Eq. 11.3-3]}$$

$$k = -1 + \sqrt{\frac{(2)(1 + R_e)}{R_e} + \frac{F_{yb}(2 + R_e)D^2}{2F_{em}t_s^2}}$$

$$R_e = \frac{F_{em}}{F_{es}}$$

$F_{em} = $ member holding point, lbf/in^2
    (See NDS Table 11A.)

F_{es} (See NDS Table 11A.)

F_{yb} = bending yield strength of wood screw, lbf/in^2

D = unthreaded shank diameter of wood screw, in

$K_D = 2.2$ for $D \leq 0.17$ in

$K_D = 10D + 0.5$ for 0.17 in $< D < 0.25$ in

$K_D = 3.0$ for $D \geq 0.25$ in

As an alternative to using these equations, NDS Tables 11.3A and 11.3B provide nominal wood screw lateral design values, Z, for most common applications. However, the equations must be used for cases not covered in these tables.

These design values, Z, are multiplied by applicable adjustment factors to obtain allowable design values, Z'.

A. Allowable Lateral Design Value, Z'
[NDS Table 7.3.1; NDS 11.3]

The allowable lateral design value is

$$Z' = ZC_DC_MC_tC_dC_{eg}$$

Z (See NDS Tables 11.3A and 11.3B.)

C_d (See NDS 11.3.3.)

$C_{eg} = 0.67$ if inserted into the end grain of main member (See NDS 11.3.4.)

Penetration Depth Factor, C_d
[NDS 11.3.3]

Since the nominal wood screw lateral design values, by equations or tables, are based on a penetration length of $7D$ (seven times the unthreaded shank diameter of the wood screw) into the main member, these values are multiplied by the penetration depth factor, C_d, if the penetration is less than $7D$ but greater than or equal to $4D$.

$$C_d = \frac{p}{7D} \leq 1.0 \qquad \text{[NDS Eq. 11.3-4]}$$

B. Placement of Wood Screws
[NDS 11.4]

Edge distances, end distances, and screw spacing shall be placed such that splitting of the wood is prevented.

For connections containing more than one wood screw, if the wood screws are of the same type and similar size and have the same yield mode, the total allowable design value is the sum of the allowable design values for each individual wood screw.

Example 10.6
Lateral Load for Wood Screws

Four no. 10 gage (0.190 in diameter, see NDS Table 11.3A) × 1.25 in wood screws and two 10-gage ($t_s =$

0.134 in) steel plates, one on each side of the connection, hold a 4×4 side member to a 4×6 main member. The lumbers are red oak that are exposed to weather in service. Assume normal temperature ($C_t = 1.0$).

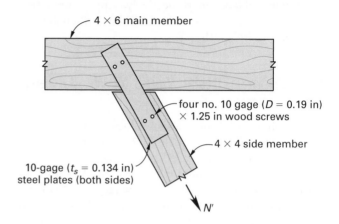

Using NDS Table 11.3, determine the allowable load, N', for a one-year load duration. Ignore the capacity of the steel plates.

Solution:

NDS Section

Table 11A $G = 0.67$ for red oak

Table 11.3B $Z = 192$ lbf

Table 7.3.3 $C_M = 0.70$ because the lumber is exposed to weather

App. B $C_D = 1.10$ for one-year load duration

C_{eg} is not applicable.

Find the penetration depth factor, C_d.

11.3.3   The actual penetration into a 4×6 or a 4×4 member is

$p = 1.25$ in $- 0.134$ in $= 1.12$ in

The minimum penetration required for full design value is
$7D = (7)(0.190$ in$) = 1.33$ in

For reduced design value,
$4D = (4)(0.190$ in$) = 0.76$ in
0.76 in < 1.12 in
$\qquad 4D < p$ [OK]

Eq. 11.3-4 $C_d = \dfrac{p}{7D} = \dfrac{1.12 \text{ in}}{1.33 \text{ in}} = 0.84$

The allowable load is
$Z' = ZC_DC_MC_tC_d$
$\quad = (192 \text{ lbf})(1.10)(0.70)(1.0)(0.84)$
$\quad = 124.2$ lbf per screw

$N' = Z'(\text{no. of wood screws})$

$\quad = \left(124.2 \; \dfrac{\text{lbf}}{\text{screw}}\right)(2 \; \text{screws})$

$\quad = 248.4 \; \text{lbf}$

(Note: NDS Eq. 11.3-5 will apply for determining allowable design values for wood screws loaded at an angle to the wood surface.) (See NDS Fig. 9A.)

Split Rings and Shear Plates

Split rings and shear plates are highly effective mechanical fasteners that can have lateral (shear) load capacities much greater than those of bolts and lag screws/bolts. Split rings come in $2^{1}/_{2}$ in and 4 in diameters, while shear plates come in $2^{5}/_{8}$ in and 4 in diameters. Their dimensions are given in NDS App. K. Design information is given in NDS Part 10.

Split rings are used for wood-to-wood connections. A split ring is fit into a groove cut into the mating surfaces of the wood members being connected. The assembly is held together by a bolt (see Fig. 11.1).

(a) wood-to-wood

Figure 11.1 Three-Member Connection with Split Rings

Shear plates can be used for wood-to-metal or wood-to-wood connections. For a three-member wood-to-metal connection, the shear plate cuts into the wood but is flush with the surface of the wood. A bolt holds together the wood, two shear plates, and two steel member plates (see Fig. 11.2b). For a three-member wood-to-wood connection, a total of four shear plates are required to connect the three wood members, and the whole connection is held together by a bolt (see Fig. 11.2a).

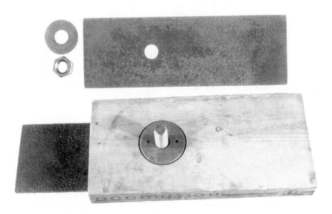

(b) wood-to-metal

Figure 11.2 Three-Member Connections with Shear Plates

1. Lumber Species Group

[NDS Table 10A]

Since the density of wood affects the lateral load capacities of split rings and shear plates, the species groupings for these connectors are based on density, as seen in Table 11.1.

Table 11.1 Species Groups for Split Ring and Shear Plate Connectors (NDS Table 10A)

group A	group B	group C	group D[a]
beech-birch-hickory	douglas fir-larch	douglas fir-south	aspen
douglas fir-larch	douglas fir-larch	eastern hemlock-	balsam fir
(dense)	(north)	tamarack (north)	coast sitka spruce
mixed oak	mixed maple	hem-fir	cottonwood
northern red oak	mixed southern pine	hem-fir (north)	eastern hemlock
red oak	red maple	mountain hemlock	eastern hemlock-
southern pine	southern pine[b]	northern pine	tamarack
(dense)		ponderosa pine	eastern softwoods
white oak		red pine	eastern spruce
		redwood (close	eastern white pine
		grain)	northern species
		sitka spruce	northern white cedar
		spruce-pine-fir	redwood (open grain)
		western hemlock	spruce-pine-fir (south)
		western hemlock	western cedars
		(north)	western cedars (north)
		yellow poplar	western white pine
			western woods

[a]Alaska cedar is in group D.

[b]Coarse grain southern pine, as used in some glued laminated timber combinations, is in group C.

Reproduced from *National Design Specification for Wood Construction*, 1997 Edition, courtesy of American Forest & Paper Association, Washington, D.C.

2. Design Values

[NDS 10.2]

NDS Tables 10.2A and 10.2B contain nominal design values parallel to grain, P, and perpendicular to grain, Q. These values are multiplied by the applicable adjustment factors specified in NDS Table 7.3.1 to obtain the respective allowable design values, P' and Q'.

$$P' = PC_DC_MC_tC_gC_\Delta C_dC_{st}$$

$$Q' = QC_DC_MC_tC_gC_\Delta C_d$$

P, Q = nominal design values
  (See NDS Tables 10.2A and 10.2B.)
C_D (See NDS 2.3.2 and 7.3.2.)
C_M (See NDS Table 7.3.3.)
C_t (See NDS Table 7.3.4.)
C_g (See NDS Tables 7.3.6B and 7.3.6D.)
C_Δ = geometry factor for edge distance, end distance, and spacing
  (See NDS 10.3.2 through 10.3.5; NDS Table 10.3.)
C_d = penetration depth factor with lag screws
  (See NDS Table 10.2.3.)
C_{st} = metal side plate factor
  (See NDS Table 10.2.4.)

Example 11.1
2¹/₂ in Single Split Rings with Parallel-to-Grain and Perpendicular-to-Grain Loadings

One 2×6 and two 2×8s are connected using $2^1/_2$ in split rings as shown (see NDS App. K for split ring dimensions). Assume that $C_D = C_M = C_t = 1.0$ and that all timbers are douglas fir-larch no. 1 grade.

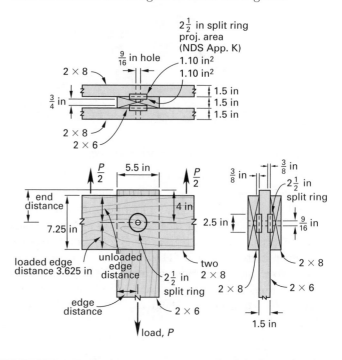

Determine the allowable load capacity for the connection.

Solution:

NDS Section

1. Analyze the allowable connector capacity for 2×6 douglas fir-larch no. 1 grade (one member having split rings parallel to grain loading).

A. The net thickness is $1\frac{1}{2}$ in, the load is parallel to the grain, and the rings are in two faces.

B. $C_D = C_M = C_t = 1.0$ (assumed in problem statement), $C_{st} = 1.0$ (no steel side plate), $C_g = 1.0$ (no group action), and $C_d = 1.0$ (no lag screws).

Table 10A   **C.** Douglas fir-larch is in species group B.

Table 10.2A P = load parallel to grain
$P = 2100$ lbf for $1\frac{1}{2}$ in minimum net member thickness.

D. Find C_Δ.

Table 10.3   Find the end distance factor for the tension members.

5.5 in for full design value; $C_\Delta = 1.0$.

2.75 in for reduced design value; $C_\Delta = 0.625$.

4.0 in for actual distance; $C_\Delta = ?$

Interpolate to find C_Δ for the actual distance of 4.0 in.
$$C_\Delta = 0.625 + \left(\frac{1.0 - 0.625}{5.5 \text{ in} - 2.75 \text{ in}} \right)$$
$$\times (4.0 \text{ in} - 2.75 \text{ in})$$
$$= 0.795 \quad \text{[controls]}$$

Table 10.3   Find the edge distance factor.

1.75 in for full design value; $C_\Delta = 1.0$.

$\dfrac{5.5 \text{ in}}{2} = 2.75$ in for actual value; $C_\Delta = ?$

Therefore, $C_\Delta = 1.0$ since actual edge distance > 1.75 in.

Table 10.3   Minimum required spacing is not applicable.

Therefore, $C_\Delta = 1.0$.

Table 7.3.1   **E.** Find the allowable load capacity per split ring.

$$P' = PC_D C_M C_t C_g C_\Delta C_d C_{st}$$
$$= (2100 \text{ lbf})(1.0)(1.0)(1.0)(1.0)(0.795)$$
$$\times (1.0)(1.0)$$
$$= 1669.5 \text{ lbf per ring}$$

Therefore, the total allowable load capacity is
$$P_{\text{allow}} = P'(\text{no. of rings})$$
$$= \left(1669.5 \ \frac{\text{lbf}}{\text{ring}} \right) (2 \text{ rings})$$
$$= 3339.0 \text{ lbf}$$

2. Analyze the allowable connector capacity for 2×8s douglas fir-larch no. 1 grade (two members having split rings perpendicular to loading).

A. The net thickness is $1\frac{1}{2}$ in, the load is perpendicular to the grain, the rings are only in one face, and the loaded edge distance is

$$\frac{7.25 \text{ in}}{2} = 3.625 \text{ in} \qquad \text{[see Fig. 9.2]}$$

Table 10A   **B.** Douglas fir-larch is in species group B.

Table 10.2A $Q = 1940$ lbf

C. Find C_Δ.

Table 10.3   Find the edge distance factor for loaded edge.

2.75 in for full design value; $C_\Delta = 1.0$.

3.625 in for actual value; $C_\Delta = ?$

Therefore, $C_\Delta = 1.0$ since actual edge distance > 2.75 in.

Table 10.3
$$Q' = QC_D C_M C_t C_g C_\Delta C_d C_{st}$$
$$= (1940 \text{ lbf})(1.0)(1.0)(1.0)(1.0)(1.0)$$
$$\times (1.0)(1.0)$$
$$= 1940 \text{ lbf per ring}$$

Therefore, the total load capacity, Q_{allow}, is
$$Q_{\text{allow}} = Q'(\text{no. of members})$$
$$= \left(1940 \ \frac{\text{lbf}}{\text{ring}} \right) (2 \text{ members})$$
$$= 3880 \text{ lbf}$$

3. Calculate the member net section load capacities.

NDS App. K   The ring's projected area is 1.10 in^2 for $2\frac{1}{2}$ in ring in a member.

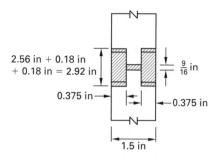

The projected area to be deducted is

$$2.92 \text{ in} \times 0.375 \text{ in} = 1.10 \text{ in}^2 \quad \left[\begin{array}{c}\text{also given in}\\ \text{NDS App. K}\end{array}\right]$$

The required bolt hole diameter is $^9/_{16}$ in for a $2^1/_2$ in ring.

A. The net areas are as follows.

The net area of 2×6 (tension load parallel to grain) is

$$A_{\text{net}} = (1.5 \text{ in})(5.5 \text{ in}) - (2)(1.10 \text{ in}^2)$$
$$- \left(\tfrac{9}{16}\text{ in}\right)\left(1.5 \text{ in} - (2)(0.375 \text{ in})\right)$$
$$= 5.63 \text{ in}^2$$

The net area of two 2×8s (tension load perpendicular to grain) is

$$A_{\text{net}} = (2)((1.5 \text{ in})(7.25 \text{ in}) - 1.10 \text{ in}^2$$
$$- \left(\tfrac{9}{16}\text{ in}\right)\left(1.5 \text{ in} - 0.375 \text{ in}\right))$$
$$= 18.28 \text{ in}^2$$

B. Find the net section load capacities.

For the 2×6 (tension load parallel to grain),

Table 2.3.1   For a 2×6 douglas fir-larch no. 1,
$$F_t' = F_t C_D C_M C_t C_F$$

Table 4A; $F_t = 675 \text{ lbf/in}^2$; $C_F = 1.3$
Adj. Factors

$$F_t' = \left(675 \frac{\text{lbf}}{\text{in}^2}\right)(1.0)(1.0)(1.0)(1.3)$$
$$= 877.5 \text{ lbf/in}^2$$

The tension load capacity parallel to grain is
$$T_\| = F_t' A_{\text{net}} = \left(877.5 \frac{\text{lbf}}{\text{in}^2}\right)(5.63 \text{ in}^2)$$
$$= 4940.3 \text{ lbf}$$

For the two 2×8s (compression load perpendicular to grain),

Table 2.3.1 $F_{c\perp}' = F_{c\perp} C_M C_t C_b$

Table 4A $F_{c\perp} = 625.0 \text{ lbf/in}^2$

Eq. 2.3-1 $$C_b = \frac{l_b + 0.375 \text{ in}}{l_b}$$
$$= \frac{2.5 \text{ in ring} + 0.375 \text{ in}}{2.5 \text{ in ring}}$$
$$= 1.15$$

$$F_{c\perp}' = \left(625.0 \frac{\text{lbf}}{\text{in}^2}\right)(1.0)(1.0)(1.15)$$
$$= 718.8 \text{ lbf/in}^2$$

App. K   For a $2^1/_2$ in split ring is $^3/_4$ in deep, the compressive dead load capacity is

$$C_{c\perp} = (2 \text{ members})(F_{c\perp}')(\text{bearing area})$$
$$= (2 \text{ members})\left(718.8 \frac{\text{lbf}}{\text{in}^2}\right)$$
$$\times ((2.5 \text{ in})(0.375 \text{ in}))$$
$$= 1347.75 \text{ lbf}$$

---

4. Summary

A. Find the connection load capacities.

$P_{\text{allow}} = 3339 \text{ lbf}$ for the 2×6

$Q_{\text{allow}} = 3880 \text{ lbf}$ for the two 2×8s

B. Find the net section load capacities.

$T_\| = 4940.3 \text{ lbf}$ for the 2×6

$C_{c\perp} = 1347.75 \text{ lbf}$ for the two 2×8s
(this value controls the design)

C. The allowable load capacity is 1347.75 lbf.

Example 11.2
Splice with Multiple 4 in Split Rings

A splice consists of two 2×8 side members and a 4×8 main member connected with 4 in split rings (see NDS App. K). The splice is fabricated wet and is dry in service. Assume $C_t = C_{di} = C_{st} = 1.0$ and that all lumber is douglas fir-larch no. 1 grade.

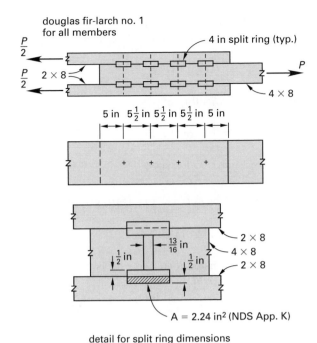

douglas fir-larch no. 1
for all members

4 in split ring (typ.)

$\frac{P}{2}$

$\frac{P}{2}$ 2 × 8

P

4 × 8

5 in 5$\frac{1}{2}$ in 5$\frac{1}{2}$ in 5$\frac{1}{2}$ in 5 in

$\frac{13}{16}$ in

2 × 8
4 × 8
2 × 8

$1\frac{1}{2}$ in

$1\frac{1}{2}$ in

A = 2.24 in² (NDS App. K)

detail for split ring dimensions

Determine the allowable snow load capacity.

Solution:

NDS Section

1. Analyze the allowable load capacity for the 4 × 8 douglas fir-larch no. 1 grade (one main member).

A. The net thickness is 3.5 in, the load is parallel to the grain, and the rings are in two faces.

B. The allowable load per split ring is

Table 2.3.1 $P' = PC_DC_MC_tC_gC_\Delta C_dC_{st}$

Table 2.3.2 $C_D = 1.15$ for snow

Table 7.3.3 $C_M = 0.80$ for dry service factor

Table 10.2.3 $C_d = 1.0$ [There is no lag screw.]

Table 10.2.4 $C_{st} = 1.0$ [There is no steel side plate.]

Table 10A **C.** Douglas fir-larch is in species group B.

Table 10.2A $P = 5260$ lbf for 3 in or thicker net member thickness (2 faces of 4 × 8 member with split rings on same bolt).

Table 10.3 **D.** Find C_Δ.

Find the end distance factor for the tension member.

7.0 in for full design value; $C_\Delta = 1.0$.

3.5 in for reduced design value; $C_\Delta = 0.625$.

5.0 in for actual distance; $C_\Delta = ?$

Interpolate to find C_Δ for 5.0 in.

$$C_\Delta = 0.625 + \left(\frac{1.0 - 0.625}{7.0 \text{ in} - 3.5 \text{ in}} \right) (5.0 \text{ in} - 3.5 \text{ in})$$
$$= 0.786$$

Find the edge distance factor.

Table 10.3 full design value = 2.75 in

actual edge distance = $\frac{7.25 \text{ in}}{2} = 3.625$ in

Therefore, $C_\Delta = 1.0$ since actual edge distance > 2.75 in.

Find the spacing factor.

Table 10.3 9.0 in for full design value; $C_\Delta = 1.0$.

5.0 in for reduced design value; $C_\Delta = 0.5$.

Interpolate to find C_Δ for the actual distance of 5.5 in.

$$C_\Delta = 0.5 + \left(\frac{1.0 - 0.5}{9 \text{ in} - 5 \text{ in}} \right) (5.5 \text{ in} - 5.0 \text{ in})$$
$$= 0.562 \quad \text{[controls]}$$

E. Find the group action factor, C_g, for four fasteners in a row.

Table 7.3.6B $A_m = (3.5 \text{ in})(7.25 \text{ in})$
$= 25.38 \text{ in}^2$ for one main member

$A_s = (2)(1.5 \text{ in})(7.25 \text{ in})$
$= 21.75 \text{ in}^2$ for two side members

$\frac{A_s}{A_m} = \frac{21.75 \text{ in}^2}{25.38 \text{ in}^2} = 0.857$

By linear interpolation, $C_g = 0.858$.

F. The allowable load capacity per split ring is

$$P' = PC_DC_MC_tC_gC_\Delta C_dC_{st}$$
$$= (5260 \text{ lbf})(1.15)(0.8)(1.0)(0.858)$$
$$\times (0.562)(1.0)(1.0)$$
$$= 2333.4 \text{ lbf per ring}$$

Therefore, the total allowable load capacity is

$$P_{\text{allow}} = \left(2333.4 \, \frac{\text{lbf}}{\text{ring}} \right) (8 \text{ rings})$$
$$= 18,667.2 \text{ lbf}$$

2. Analyze the allowable load capacity for the two 2 × 8s douglas fir-larch no. 1 grade (two side members).

A. The net thickness is $1^{1}/_{2}$ in, the load is parallel to the grain, and the rings are only in one face.

Table 10A   **B.** Douglas fir-larch is in species group B.

Table 2.3.1 $P' = P C_D C_M C_t C_g C_\Delta C_d C_{st}$
(Adjustment factors were determined previously.)

Table 10.2A $P = 5160$ lbf/ring

$$P' = \left(5160 \; \frac{\text{lbf}}{\text{ring}}\right)(1.15)(0.8)(1.0)(0.858)$$
$$\times (0.562)(1.0)(1.0)$$
$$= 2289.1 \; \text{lbf/ring}$$

Therefore, the total allowable load capacity is

$$P_{\text{allow}} = \left(2289.1 \; \frac{\text{lbf}}{\text{ring}}\right)(4 \; \text{rings})$$
$$\times (2 \; \text{members})$$
$$= 18{,}312.9 \; \text{lbf}$$

---

3. Analyze the member net section load capacities.

NDS App. K   The projected area of 4 in ring is 2.24 in^2 per member. The split ring depth is 1 in.

The required bolt hole diameter is $^{13}/_{16}$ in for a 4 in split ring.

A. The net areas are as follows.

For the 4 × 8,
$$A_{\text{net}} = (3.5 \; \text{in})(7.25 \; \text{in}) - (2)(2.24 \; \text{in}^2)$$
$$- \left(\tfrac{13}{16} \; \text{in}\right)(3.5 \; \text{in} - (2)(0.50 \; \text{in}))$$
$$= 18.86 \; \text{in}^2$$

For the two 2 × 8s,
$$A_{\text{net}} = (2)\big((1.5 \; \text{in})(7.25 \; \text{in}) - 2.24 \; \text{in}^2$$
$$- \left(\tfrac{13}{16} \; \text{in}\right)(1.5 \; \text{in} - 0.5 \; \text{in})\big)$$
$$= 15.64 \; \text{in}^2 \quad \left[\begin{array}{l}\text{controls since this is}\\\text{the least net area}\end{array}\right]$$

B. The net section load capacity parallel to the grain for the two 2 × 8s is

Table 2.3.1 $F_t' = F_t C_D C_M C_t C_F$

Table 4A $F_t = 675 \; \text{lbf/in}^2$

$C_F = 1.2$ for 2 × 8

$C_D = 1.15$ for snow

$$F_t' = \left(675.0 \; \frac{\text{lbf}}{\text{in}^2}\right)(1.15)(0.8)(1.0)(1.2)$$
$$= 745.2 \; \text{lbf/in}^2$$

The tension load capacity is
$$P_{\text{allow}} = F_t' A_{\text{net}}$$
$$= \left(745.2 \; \frac{\text{lbf}}{\text{in}^2}\right)(15.64 \; \text{in}^2)$$
$$= 11{,}654.9 \; \text{lbf}$$
$$\text{for two 2 × 8 side members}$$

---

4. Summary

A. Find the connection capacities.

$P_{\text{allow}} = 18{,}667.2$ lbf for the 4×8 main member.

$P_{\text{allow}} = 13{,}308.8$ lbf for the two 2 × 8 side members.

B. Find the net section load capacities.

$P_{\text{allow}} = 11{,}654.9$ lbf for the two 2 × 8 side members.

C. The allowable snow load capacity is the lowest capacity: 11,654.9 lbf.

Example 11.3
4 in Split Rings with Lag Screws

A truss connection consists of a 4 in split ring and a $^3/_4$ in by 6 in long lag screw as shown. The wood is seasoned douglas fir-larch no. 1 grade that remains wet in service.

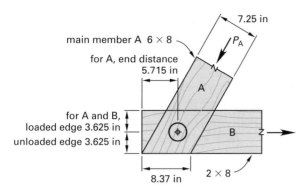

(a) member A

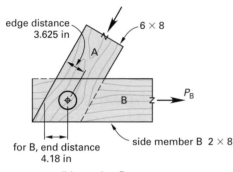

(b) member B

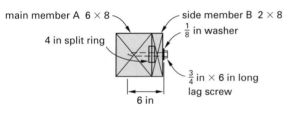

(c) cross section

Determine the wind load capacity, P_{allow}, for the truss connection.

Solution:

NDS Section

Table 10A — Douglas fir-larch is in species group B.

The allowable design value is

Table 7.3.1 $P' = PC_D C_M C_t C_g C_\Delta C_d C_{st}$

Table 7.3.3 $C_M = 0.70$ for wet in service

Table 2.3.2 $C_D = 1.6$ for wind loads

Table 2.3.4 $C_t = 1.0$ for normal temperature

Table 7.3.6B $C_g = 1.0$ for one split ring

10.2.4 $C_{st} = 1.0$ no steel side plates

Table 10.2.3 — Find the penetration factor into main member, C_d.

The minimum required penetration for full design value, $C_d = 1.0$, is

$8D = (8)\left(\frac{3}{4}\text{ in}\right) = 6\text{ in}$

The minimum required penetration for reduced design value, $C_d = 0.75$, is

$3.5D = (3.5)\left(\frac{3}{4}\text{ in}\right) = 2.625\text{ in}$

App. L — The actual penetration into the 6×8 main member is

$6\text{ in} - 1.5\text{ in for side member}$
$\quad - 0.125\text{ in for washer} - 0.5\text{ in for } E$
$\quad = 3.875\text{ in}$

Table 10.2.3 $C_d = 0.75 + \left(\dfrac{1.0 - 0.75}{6\text{ in} - 2.625\text{ in}}\right)$
$\qquad \times (3.875\text{ in} - 2.625\text{ in})$
$\qquad = 0.843$

---

1. Analyze the allowable connector capacity for the 2×8 side member (member B).

A. The net thickness is $1\frac{1}{2}$ in, the load is parallel to the grain, the ring is only in one face, the loaded edge distance is 3.625 in, and douglas fir-larch is in species group B.

Table 10.2A **B.** $P = 5160.0$ lbf

C. Find C_Δ.

Table 10.3 — For the 6×8 main member in compression (member A), find the edge distance factor.

2.75 in minimum edge distance for full design value; $C_\Delta = 1.0$. 3.625 in is the actual value for a loaded or unloaded edge. Therefore, $C_\Delta = 1.0$ since actual edge distance > 2.75 in.

Table 10.3 — Find the end distance factor for compression member.

5.5 in minimum end distance for full design value; $C_\Delta = 1.0$. 5.715 in is the actual value. Therefore, $C_\Delta = 1.0$ since the actual end distance > 5.5 in.

Spacing factor is not applicable. Therefore, $C_\Delta = 1.0$.

Table 10.3 — For the 2×8 side member in tension (member B), find the edge distance factor.

2.75 in for full design value; $C_\Delta = 1.0$. 3.625 in is the actual value. Therefore, $C_\Delta = 1.0$ since actual edge distance > 2.75 in.

Table 10.3 — The end distance for the tension member is 7.0 in for full design value: $C_\Delta = 1.0$; 3.5 in for reduced design value: $C_\Delta = 0.625$; and 4.18 in for actual distance: $C_\Delta = ?$

Interpolate to find C_Δ for the actual distance of 4.18 in.

$$C_\Delta = 0.625 + \left(\frac{1.0 - 0.625}{7 \text{ in} - 3.5 \text{ in}} \right) (4.18 \text{ in} - 3.50 \text{ in})$$

$$= 0.698$$

Find the allowable wind load capacity of the connector.

$$P'_B = P_{\text{allow}} = P_B C_D C_M C_g C_\Delta C_d C_{st}$$
$$= (5160.0 \text{ lbf})(1.60)(0.70)(1.0)(0.698)$$
$$\times (0.843)(1.0)$$
$$= 3400.5 \text{ lbf}$$

---

2. Find the net section load capacity of the 2×8 in tension.

App. K

For a 4 in split ring, the projected area is 2.24 in^2 in a member.

The required bolt hole diameter is $^{13}/_{16}$ in for a 4 in split ring.

A. The net area of one 2×8 is

$$A_{\text{net}} = (1.5 \text{ in})(7.25 \text{ in}) - 2.24 \text{ in}^2$$
$$- \left(\tfrac{13}{16} \text{ in} \right) (1.5 \text{ in} - 0.5 \text{ in})$$

$$= 7.83 \text{ in}^2$$

B. The net section load capacity, $T_{\|}$, is found as follows.

Table 2.3.1 $F'_t = F_t C_D C_M C_t C_F$

Table 4A; $F_t = 675 \text{ lbf/in}^2$

Adj. Factors $C_F = 1.2$ for 2×8

$C_D = 1.6$

$C_M = 1.0$

$$F'_t = \left(675 \ \frac{\text{lbf}}{\text{in}^2} \right) (1.6)(1.0)(1.0)(1.2)$$
$$= 1296 \text{ lbf/in}^2$$

The tension load capacity is
$$P_{\text{allow}} = T_{\|} = F'_t A_{\text{net}}$$
$$= \left(1296 \ \frac{\text{lbf}}{\text{in}^2} \right) (7.82 \text{ in}^2)$$
$$= 10{,}134.7 \text{ lbf}$$

C. The allowable wind load capacity by net section ($P_{\text{allow}} = T_{\|}$) is 10,134.7 lbf.

---

3. The allowable wind load capacity for the connection, P_{allow}, is controlled by the connector capacity and is 3400.5 lbf.

Plywood and Nonplywood Structural Panels

1. Introduction

Plywood has many varied structural applications. It is valuable because of its size and physical properties. A piece of solid-sawn wood $1/2$ in $\times$ 48 in $\times$ 96 in would be very expensive and structurally useless; however, the same size piece of plywood would be much less expensive and, in some ways, stronger than the wood from which it was cut.

Virtually any softwood species can be used in plywood. Plywood is made by peeling logs into veneer and laminating layers of this veneer with glue. The strength results from cutting up and spreading out the knots and other defects, and cross-banding the laminae—that is, placing alternate laminae perpendicular to each other. The result is that plywood has two strong directions instead of one, as in solid timber.

A. Plywood Grades and Types

Plywood is available in many forms and grades. The variables in plywood composition are veneer grade and species, veneer configuration, and glue type. Structural plywood veneers are graded for quality from A to D, A being the highest grade. Plywood is designated by the veneer grades that appear in the outer, or face plies. C-D plywood, for example, has a C grade front face with smaller knot holes and a D grade veneer on the back face.

With all the potential variations in plywood composition, it is only through the standards established by the American Plywood Association that the end users can know what to expect from a given sheet.[1] The performance-based standards are written as fabrication limits for equivalent plywood designations.

The wood species group system (Table 12.1) is to simplify the design and identification of a plywood panel. The group classification of a panel identifies the face and

---

[1] American Plywood Association, 7011 South 19th St., Tacoma, Washington 98411-0700.

back veneers, and the inner veneers can be of a different group. Marine and structural I grades, however, are required to have all plies of group 1 species. Group 1 consists of the strongest varieties of wood, and group 4 comprises the weakest varieties of wood.

There are four durability classifications: exterior (permanently exposed to the weather), exposure 1 (not permanently exposed to the weather), IMG or exposure 2 (protected applications that are not continuously exposed to high humidity conditions), and interior (permanently protected interior applications).

B. Plywood Structural Applications

The most common plywood applications are flooring, roofing, and siding. The plywood spans the space between joists, rafters, or studs and distributes loads to those members. Engineering calculations can be used to investigate the plywood stresses under these conditions, but it is much simpler to follow the allowable span recommendations found in the certification stamp for most common decking plywoods. Figure 12.1 is an example of a plywood stamp. The 32/16 is the span rating. A span rating of 32/16 indicates that the panel can be used to span 32 in in a roof system and 16 in as floor sheathing.

The span rating for sheathing panels is a measure of the panel stiffness and strength parallel to the face grain. The span rating is used for a roof or floor sheathing without engineering design calculations (Table 12.2). The span rating consists of two numbers separated by a slash. The number on the left in the span rating gives the maximum spacing for roof supports under average loading conditions (good for a live load of 30 lbf/ft^2 or better). The number to the right of the slash shows the maximum spacing for floor supports under average residential loading (maximum allowable uniform loads are 100 lbf/ft^2 or more, depending upon the allowable deflections). It is important to note that the span is in the face-grain direction, which is parallel to the 8 ft edge of a standard 4 ft $\times$ 8 ft plywood panel.

Table 12.1 Classification of Species (APA PDS Table 1.5)

group 1	group 2		group 3	group 4	group 5[a]
Apitong[b,c]	cedar, Port Orford	maple, black	alder, red	aspen	basswood
beech, American	cypress	Mengkulang[b]	birch, paper	bigtooth	poplar, balsam
birch	douglas-fir 2[d]	Meranti, red[b,e]	cedar, Alaska	quaking	
sweet	fir	Mersawa[b]	fir, subalpine	Cativo	
yellow	balsam	pine	hemlock, eastern	cedar	
douglas-fir 1[d]	California red	pond	maple, bigleaf	incense	
Kapur[b]	grand	red	pine	western red	
Keruing[b,c]	noble	Virginia	Jack	cottonwood	
larch, western	Pacific silver	western white	lodgepole	eastern	
maple, sugar	white	spruce	ponderosa	black (western	
pine	hemlock, western	black	spruce	poplar)	
Caribbean	lauan	red	redwood	pine	
Ocote	Almon	Sitka	spruce	eastern white	
pine, southern	Bagtikan	sweetgum	Engelmann	sugar	
loblolly	Mayapis	tamarack	white		
longleaf	red	yellow-poplar			
shortleaf	Tangile				
slash	white				
tan oak					

[a]Design stresses for group 5 not assigned.

[b]Each of these names represents a trade group of woods consisting of a number of closely related species.

[c]Species from the genus Dipterocarpus are marketed collectively: Apitong if originating in the Philippines; Keruing if originating in Malaysia or Indonesia.

[d]Douglas-fir from trees grown in the states of Washington, Oregon, California, Idaho, Montana, Wyoming, and the Canadian provinces of Alberta and British Columbia shall be classed as douglas-fir no. 1. Douglas-fir from trees grown in the states of Nevada, Utah, Colorado, Arizona and New Mexico shall be classed as douglas-fir no. 2.

[e]Red Meranti shall be limited to species having a specific gravity of 0.41 or more based on green volume and oven dry weight.

Used with permission of APA, the Engineered Wood Association, Tacoma, WA.

Table 12.2 Key to Span Rating and Species Group (APA PDS)

For panels with "Span Rating" as across top, and thickness as at left, use stress for species group given in table.

thickness (in)	span rating (APA rated sheathing grades)									
	12/0	16/0	20/0	24/0	32/16	40/20	48/24			
					span rating (sturd-I-floor grades)					
					16 OC	20 OC	24 OC	48 OC		
5/16	4	3	1							
3/8			4	1[a]						
15/32 and 1/2				4	1					
19/32 and 5/8					4	1				
23/32 and 3/4						4	1			
7/8							3[b]			
1 1/8								1		

[a]Thicknesses not applicable to APA rated STURD-I-FLOOR.
[b]For APA rated STURD-I-FLOOR 24 OC, use group 4 stresses.

Used with permission of APA, the Engineered Wood Association, Tacoma, WA.

Plywood sheathing on walls, roofs, and floors can resist more than just loads normal to the surface. If the plywood is sized and connected adequately, the walls, roofs, and floors can act as shear diaphragms in resisting lateral loads on a building.

Plywood is very strong and rigid against forces that tend to distort it out of square. Its strength and shape make plywood a logical choice for use as the web of a built-up beam. Box and I-beams can be fabricated with lumber flanges.

It is also possible to make an efficient double use of plywood sheathing in stressed-skin or sandwich panels. In these structural elements, the plywood skin acts as the flanges of a shallow, wide beam. Shear resistance is provided by the lumber webs or sandwich core material. Supplements to the *Plywood Design Specification*, published by the APA, provide the information needed to design and analyze these relatively sophisticated elements.

2. Plywood Section Properties

A. Direction of Face Grain

In plywood, alternating plies are usually perpendicular. Since wood has markedly different properties across and along the grain, the direction in which the plies are oriented is important. The standard orientation reference is the grain direction of the visible face plies. The grain in the face plies of a 4 ft × 8 ft plywood panel almost always runs in the 8 ft direction.

The effective section properties of Table 12.3 depend on the orientation of the stresses relative to this face-grain direction. Plywood used as sheathing is strongest

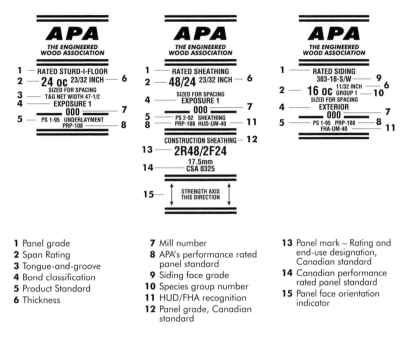

1 Panel grade
2 Span Rating
3 Tongue-and-groove
4 Bond classification
5 Product Standard
6 Thickness

7 Mill number
8 APA's performance rated panel standard
9 Siding face grade
10 Species group number
11 HUD/FHA recognition
12 Panel grade, Canadian standard

13 Panel mark – Rating and end-use designation, Canadian standard
14 Canadian performance rated panel standard
15 Panel face orientation indicator

Figure 12.1 APA Grade-Trademark Stamp

Used with permission of APA, the Engineered Wood Association, Tacoma, WA.

against normal loads if the face grain is across the supporting members. Plywood loaded in this strong direction is an example of the Stress Applied Parallel to Face Grain category used in Table 12.3 (*Plywood Design Specification*, 1997).

B. Thickness for All Properties Except Shear

For all calculations other than those involving shear, the nominal thickness found in column 1 of Table 12.3 is used.

C. Thickness for Shear

For calculating shear stresses, the effective thicknesses of column 3 in Table 12.3 are used. For structural plywood grades, this effective thickness can be larger than the actual, nominal thickness.

D. Cross-Sectional Area

The effective areas of columns 4 and 8 in Table 12.3 reflect the differences in grain orientation. Those plies with fibers perpendicular to the direction of stress application are neglected as having essentially no stiffness or strength. Note that the effective areas are much larger when the stress is applied parallel to the face.

The effective area is given in units of square inches per foot. This per-foot basis treats the membrane of plywood as a series of interconnected, 1 ft wide beams and is common to the rest of the section properties. The foot represents a 12 in wide strip of the specific piece of plywood, measured perpendicular to the direction of stress application. A 4 ft × 8 ft panel with the stress

parallel to the face grain would, therefore, have an effective area resisting that stress of four times the value found in column 4 of Table 12.3.

E. Moment of Inertia

Only those plies whose fibers are parallel to the stress direction are included in the values of columns 5 and 9 of Table 12.3. The per-foot unit means the same as it does for effective area.

F. Section Modulus

The effective section modulus, KS, includes the empirical factor, K. This KS value should always be used for bending stress calculations when the plywood is installed as sheathing and loaded normal to the surface. Specifically, the apparently valid I/c value for an effective section modulus should not be used. Figure 12.2 illustrates the difference between loading in the plane of the plywood and loading normal to that plane.

When plywood is used as the web of a plywood-lumber beam, the loads are in the plane of the panel and bending stresses in the plywood are more like axial forces. A different section modulus will be used in those calculations.

G. Rolling Shear Constant

When plywood is loaded in shear through the thickness (see Fig. 12.3), it is tremendously strong because the cross plies are being sheared across the grain. When the shear stresses lie in the plane of the plies, however, plywood is not nearly as strong. This rolling shear that develops (see Fig. 12.3) tends to roll the fibers of the cross plies over each other, a tendency against which wood has little resistance. Rolling shear stresses arise in plywood used as sheathing and at the connection between plywood webs and lumber flanges in plywood-lumber beams.

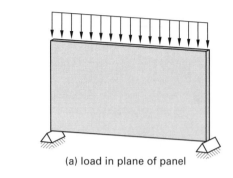

(a) load in plane of panel

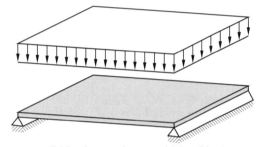

(b) load normal to panel (sheathing)

Figure 12.2 Plywood in Bending

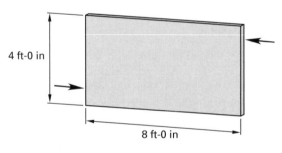

4 ft-0 in

8 ft-0 in

shear through thickness

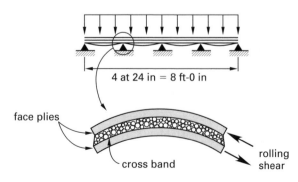

4 at 24 in = 8 ft-0 in

face plies

cross band

rolling shear

Figure 12.3 Shear Stress Orientations

3. Allowable Stresses: Plywood

There are two basic factors that determine the allowable stresses in plywood: the species used in the laminae and the method of assembling those laminae. The allowable stresses are further subjected to modification for load duration and service moisture conditions.

A. Grade Stress Levels

The first distinction drawn in the allowable plywood stress tables is the grades of the laminae used in the plywood. Plywood is classified into three grade stress levels: S-1, S-2, and S-3. The highest grade stress level, S-1, is associated with plywood that uses only exterior glue and top-grade veneers. The lowest grade stress level, S-3, is reserved for plywood with interior glue. The intermediate grade stress level, S-2, incorporates all the other plywoods. Table 12.4 lists the most common structural plywood types and gives the grade stress level by plywood type.

B. Plywood Species Groups and Span Rating

The next distinction required to determine the allowable stress is the species used to make the laminae. The seventy-plus wood species commonly used in plywood are divided into five groups.

If the species is known, Table 12.1 provides the plywood species classification information.

Often, however, the plywood manufacturer will not stamp the plywood with the species. The stamp could include an allowable roof/floor span rating that can be used to indirectly determine the species. Table 12.2 gives the species group as a function of the plywood thickness and span rating. For example, a $^5/_8$ in piece of plywood with group 4 species plies and a $^1/_2$ in sheet of group 1 species plywood will both handle 32 in rafter spacings and 16 in joist spacings.

C. Service Moisture Conditions

There are only two conditions of service that apply to plywood allowable stresses: wet and dry. If the equilibrium moisture content will be less than 16% in service, the dry use values should be used. As long as the plywood is not directly exposed to the weather, this dry condition can be assumed.

D. Allowable Stress Tables

Table 12.5 lists the allowable stresses and stiffnesses that should be used in plywood design.

E. Duration of Load Factors

Plywood is a wood product, so its load capacity is time-dependent. The load duration factors of NDS Table 2.3.2 also apply to the allowable plywood stresses of Table 12.5.

Table 12.3 Effective Section Properties for Plywood (APA PDS Tables 1 and 2)

APA PDS TABLE 1. Face Plies of Different Species Group from Inner Plies (includes all product standard grades except those noted in Table 2.)

nominal thickness (in)	approximate weight (lbf/ft²)	t_s effective thickness for shear (in)	stress applied parallel to face grain				stress applied perpendicular to face grain			
			A area (in²/ft)	I moment of inertia (in⁴/ft)	KS effective section modulus (in³/ft)	Ib/Q rolling shear constant (in²/ft)	A area (in²/ft)	I moment of inertia (in⁴/ft)	KS effective section modulus (in³/ft)	Ib/Q rolling shear constant (in²/ft)
unsanded panels										
5/16 -U	1.0	0.268	1.491	0.022	0.112	2.569	0.660	0.001	0.023	4.497
3/8 -U	1.1	0.278	1.866	0.039	0.152	3.110	0.799	0.002	0.033	5.444
15/32 and 1/2 -U	1.5	0.298	2.292	0.067	0.213	3.921	1.007	0.004	0.056	2.450
19/32 and 5/8 -U	1.8	0.319	2.330	0.121	0.379	5.004	1.285	0.010	0.091	3.106
23/32 and 3/4 -U	2.2	0.445	3.247	0.234	0.496	6.455	1.563	0.036	0.232	3.613
7/8 -U	2.6	0.607	3.509	0.340	0.678	7.175	1.950	0.112	0.397	4.791
1 -U	3.0	0.842	3.916	0.493	0.859	9.244	3.145	0.210	0.660	6.533
1 1/8 -U	3.3	0.859	4.725	0.676	1.047	9.960	3.079	0.288	0.768	7.931
sanded panels										
1/4 -S	0.8	0.267	0.996	0.008	0.059	2.010	0.348	0.001	0.009	2.019
11/32 -S	1.0	0.284	0.996	0.019	0.093	2.765	0.417	0.001	0.016	2.589
3/8 -S	1.1	0.288	1.307	0.027	0.125	3.088	0.626	0.002	0.023	3.510
15/32 -S	1.4	0.421	1.947	0.066	0.214	4.113	1.204	0.006	0.067	2.434
1/2 -S	1.5	0.425	1.947	0.077	0.236	4.466	1.240	0.009	0.087	2.752
19/32 -S	1.7	0.546	2.423	0.115	0.315	5.471	1.389	0.021	0.137	2.861
5/8 -S	1.8	0.550	2.475	0.129	0.339	5.824	1.528	0.027	0.164	3.119
23/32 -S	2.1	0.563	2.822	0.179	0.389	6.581	1.737	0.050	0.231	3.818
3/4 -S	2.2	0.568	2.884	0.197	0.412	6.762	2.081	0.063	0.285	4.079
7/8 -S	2.6	0.586	2.942	0.278	0.515	8.050	2.651	0.104	0.394	5.078
1 -S	3.0	0.817	3.721	0.423	0.664	8.882	3.163	0.185	0.591	7.031
1 1/8 -S	3.3	0.836	3.854	0.548	0.820	9.883	3.180	0.271	0.744	8.428
touch-sanded panels										
1/2 -T	1.5	0.342	2.698	0.083	0.271	4.252	1.159	0.006	0.061	2.746
19/32 and 5/8 -T	1.8	0.408	2.354	0.123	0.327	5.346	1.555	0.016	0.135	3.220
23/32 and 3/4 -T	2.2	0.439	2.715	0.193	0.398	6.589	1.622	0.032	0.219	3.635
1 1/8 -T	3.3	0.839	4.548	0.633	0.977	11.258	4.067	0.272	0.743	8.535

APA PDS TABLE 2. STRUCTURAL I and Marine

nominal thickness (in)	approximate weight (lbf/ft²)	t_s effective thickness for shear (in)	stress applied parallel to face grain				stress applied perpendicular to face grain			
			A area (in²/ft)	I moment of inertia (in⁴/ft)	KS effective section modulus (in³/ft)	Ib/Q rolling shear constant (in²/ft)	A area (in²/ft)	I moment of inertia (in⁴/ft)	KS effective section modulus (in³/ft)	Ib/Q rolling shear constant (in²/ft)
unsanded panels										
5/16 -U	1.0	0.356	1.619	0.022	0.126	2.567	1.188	0.002	0.029	6.037
3/8 -U	1.1	0.371	2.226	0.041	0.195	3.107	1.438	0.003	0.043	7.307
15/32 and 1/2 -U	1.5	0.535	2.719	0.074	0.279	4.157	2.175	0.012	0.116	2.408
19/32 and 5/8 -U	1.8	0.707	3.464	0.154	0.437	5.685	2.742	0.045	0.240	3.072
23/32 and 3/4 -U	2.2	0.739	4.219	0.236	0.549	6.148	2.813	0.064	0.299	3.540
7/8 -U	2.6	0.776	4.388	0.346	0.690	6.948	3.510	0.131	0.457	4.722
1 -U	3.0	1.088	5.200	0.529	0.922	8.512	5.661	0.270	0.781	6.435
1 1/8 -U	3.3	1.118	6.654	0.751	1.164	9.061	5.542	0.408	0.999	7.833
sanded panels										
1/4 -S	0.8	0.342	1.280	0.012	0.083	2.009	0.626	0.001	0.013	2.723
11/32 -S	1.0	0.365	1.280	0.026	0.133	2.764	0.751	0.001	0.023	3.397
3/8 -S	1.1	0.373	1.680	0.038	0.177	3.086	1.126	0.002	0.033	4.927
15/32 -S	1.4	0.537	1.947	0.067	0.246	4.107	2.168	0.009	0.093	2.405
1/2 -S	1.5	0.545	1.947	0.078	0.271	4.457	2.232	0.014	0.123	2.725
19/32 -S	1.7	0.709	3.018	0.116	0.338	5.566	2.501	0.034	0.199	2.811
5/8 -S	1.8	0.717	3.112	0.131	0.361	5.934	2.751	0.045	0.238	3.073
23/32 -S	2.1	0.741	3.735	0.183	0.439	6.109	3.126	0.085	0.338	3.780
3/4 -S	2.2	0.748	3.848	0.202	0.464	6.189	3.745	0.108	0.418	4.047
7/8 -S	2.6	0.778	3.952	0.288	0.569	7.539	4.772	0.179	0.579	5.046
1 -S	3.0	1.091	5.215	0.479	0.827	7.978	5.693	0.321	0.870	6.981
1 1/8 -S	3.3	1.121	5.593	0.623	0.955	8.841	5.724	0.474	1.098	8.377
touch-sanded panels										
1/2 -T	1.5	0.543	2.698	0.084	0.282	4.511	2.486	0.020	0.162	2.720
19/32 and 5/8 -T	1.8	0.707	3.127	0.124	0.349	5.500	2.799	0.050	0.259	3.183
23/32 and 3/4 -T	2.2	0.739	4.059	0.201	0.469	6.592	3.625	0.078	0.350	3.596

Table 12.4a Guide to Use of Allowable Stress and Section Properties Tables

INTERIOR OR PROTECTED APPLICATIONS

Plywood Grade	Description and Use	Typical Trademarks	Veneer Grade			Common Thicknesses	Grade Stress Level (Table 3)	Species Group	Section Property Table
			Face	Back	Inner				
APA RATED SHEATHING EXP 1 or 2[3]	Unsanded sheathing grade for wall, roof, suflooring, and industrial applications such as pallets and for engineering design, with proper stresses. Manufactured with intermediate and exterior glue.(1) For permanent exposure to weather or moisture only Exterior type plywood is suitable.	APA THE ENGINEERED WOOD ASSOCIATION RATED SHEATHING 32/16 15/32 INCH SIZED FOR SPACING EXPOSURE 1 000 PS 1-95 C-D PRP-108	C	D	D	5/16, 3/8, 15/32, 1/2, 19/32, 5/8, 23/32, 3/4	S-3[1]	See "Key to Span Rating"	Table 12.3 (PDS Table 1) (unsanded)
APA STRUCTURAL I RATED SHEATHING EXP 1[3]	Plywood grades to use where shear and cross-panel strength propeties are of maximum importance. Made with exterior glue only. Structural I is made from all Group 1 woods.	APA THE ENGINEERED WOOD ASSOCIATION RATED SHEATHING STRUCTURAL I 24/0 3/8 INCH SIZED FOR SPACING EXPOSURE 1 000 PS 1-95 C-D PRP-108	C	D	D	5/16, 3/8, 15/32, 1/2, 19/32, 5/8, 23/32, 3/4	S-2	Group 1	Table 12.3 (PDS Table 2) (unsanded)
APA RATED STURD-I-FLOOR EXP 1 or 2[3]	For combination subfloor-underlayment. Provides smooth surface for application of carpet and pad. Possesses high concentrated and impact load resitance during construction and occupancy. Manufactured with intermediate and exterior glue. Touch-sanded.(4) Available with tongue-and-groove edges.(5)	APA THE ENGINEERED WOOD ASSOCIATION RATED STURD-I-FLOOR 20 oc 19/32 INCH SIZED FOR SPACING T&G NET WIDTH 47-1/2 EXPOSURE 1 000 PS 1-95 UNDERLAYMENT PRP-108	C plugged	D	C & D	19/32, 5/8, 23/32, 3/4, 1-1/8 (2-4-1)	S-3[1]	See "Key to Span Rating"	Table 12.3 (PDS Table 1) (touch-sanded)
APA UNDERLAYMENT EXP 1, 2 or INT	For underlayment under carpet and pad. Available with exterior glue. Touch-sanded. Available with tongue-and-groove edges.(5)	APA THE ENGINEERED WOOD ASSOCIATION UNDERLAYMENT GROUP 1 EXPOSURE 1 000 PS 1-95	C plugged	D	C & D	1/2, 19/32, 5/8, 23/32, 3/4	S-3[1]	As specified	Table 12.3 (PDS Table 1) (touch-sanded)
APA C-D PLUGGED EXP 1, 2 or INT	For built-ins, wall and ceiling tile backing, Not for underlayment. Available with exterior glue. Touch-sanded.(5)	APA THE ENGINEERED WOOD ASSOCIATION C-D PLUGGED GROUP 2 EXPOSURE 1 000 PS 1-95	C plugged	D	D	1/2, 19/32, 5/8, 23/32, 3/4	S-3[1]	As Specified	Table 12.3 (PDS Table 1) (touch-sanded)
APA APPEARANCE GRADES EXP 1, 2 or INT	Generally applied where a high quality surface is required. Includes APA N-N, N-A, N-B, N-D, A-A, A-B, A-D, B-B, and B-D INT grades.(5)	APA THE ENGINEERED WOOD ASSOCIATION A-D GROUP 1 EXPOSURE 1 000 PS 1-95	B or better	D or better	C & D	1/4, 11/32, 3/8, 15/32, 1/2, 19/32, 5/8, 23/32, 3/4	S-3[1]	As Specified	Table 12.3 (PDS Table 1) (sanded)

(1) When exterior glue is specified, i.e. Exposure 1, stress level 2 (S-2) should be used.

(2) Check local suppliers for availability before specifying Plyform Class II grade, as it is rarely manufactured.

(3) Properties and stresses apply only to APA RATED STURD-I-FLOOR and APA RATED SHEATHING manufactured entirely with veneers.

(4) APA RATED STURD-I-FLOOR 2-4-1 may be produced unsanded.

(5) May be available as Structural I. For such designation use Group 1 stresses and Table 2 section properties.

(6) C face and back must be natural unrepaired; if repaired, use stress level 2 (S-2).

Used with permission of APA, the Engineered Wood Association, Tacoma, WA.

The allowable stresses of Table 12.5 apply to plywood panels that are at least 24 in wide. If the plywood is used in narrower strips, there is an increased possibility of a defect appearing in a critical section. Allowable stresses should be linearly decreased from full strength at 24 in wide to half strength at 8 in wide.

Example 12.1
Plywood Properties and Allowable Stresses

Roof joists are 24 in on centers, and the roof snow load is expected to be 25 lbf/ft^2. $^3/_8$ in APA rated sheathing, exposure 1, 4 ft by 8 ft panels are turned in the strong direction; that is, the face grain is parallel to the 24 in span.

Table 12.4b Guide to Use of Allowable Stress and Section Properties Tables

EXTERIOR APPLICATIONS

Plywood Grade	Description and Use	Typical Trademarks	Veneer Grade			Common Thicknesses	Grade Stress Level (Table 3)	Species Group	Section Property Table
			Face	Back	Inner				
APA RATED SHEATHING EXT[3]	Unsanded sheathing grade with waterproof glue bond for wall, roof, subfloor and industrial applications such as pallet bins.	**APA** THE ENGINEERED WOOD ASSOCIATION RATED SHEATHING 48/24 23/32 INCH SIZED FOR SPACING EXTERIOR 000 PS 1-95 C-C PRP-108	C	C	C	5/16, 3/8, 15/32, 1/2, 19/32, 5/8 23/32, 3/4	S-1[6]	See "Key to Span Rating"	Table 12.3 (PDS Table 1) (unsanded)
APA STRUCTURAL I RATED SHEATHING EXT[3]	"Structural" is a modifier for this unsanded sheathing grade. For engineered applications in construction and industry where full Exterior-type panels are required. Structural I is made from Group 1 woods only.	**APA** THE ENGINEERED WOOD ASSOCIATION RATED SHEATHING STRUCTURAL I 24/0 3/8 INCH SIZED FOR SPACING EXPOSURE 1 000 PS 1-95 C-D PRP-108	C	C	C	5/16, 3/8, 15/32, 1/2, 19/32, 5/8, 23/32, 3/4	S-1[6]	Group 1	Table 12.3 (PDS Table 2) (unsanded)
APA RATED STURD-I-FLOOR EXT[3]	For combination subfloor-underlayment where severe moisture conditions may be present, as in balcony decks. Possesses high concentrated and impact load resistance during construction and occupancy. Touch-sanded.(4) Available with tongue-and-groove edges.(5)	**APA** THE ENGINEERED WOOD ASSOCIATION RATED STURD-I-FLOOR 20 oc 19/32 INCH SIZED FOR SPACING EXTERIOR 000 PS 1-95 C-C PLUGGED PRP-108	C plugged	C	C	19/32, 5/8, 23/32, 3/4	S-2	See "Key to Span Rating"	Table 12.3 (PDS Table 1) (touch-sanded)
APA UNDERLAYMENT EXT and APA C-C-PLUGGED EXT	Underlayment for floor where severe moisture conditions may exist. Also for controlled atmosphere rooms and many industrial applications. Touch-sanded. Available with tongue-and-groove edges.(5)	**APA** THE ENGINEERED WOOD ASSOCIATION C-C PLUGGED GROUP 2 EXTERIOR 000 PS 1-95	C plugged	C	C	1/2, 19/32, 5/8, 23/32, 3/4	S-2	As Specified	Table 12.3 (PDS Table 1) (touch-sanded)
APA B-B PLYFORM CLASS I or II[2]	Concrete-form grade with high reuse factor. Sanded both sides, mill-oiled unless otherwise specified. Available in HDO. For refined design information on this special-use panel see *APA Design/Construction Guide: Concrete Forming,* Form No. V345. Design using values from this specification will result in a conservative design.(5)	**APA** THE ENGINEERED WOOD ASSOCIATION PLYFORM B-B CLASS 1 EXTERIOR 000 PS 1-95	B	B	C	19/32, 5/8, 23/32, 3/4	S-2	Class I use Group 1; Class II use Group 3	Table 12.3 (PDS Table 1) (sanded)
APA MARINE EXT	Superior Exterior-type plywood made only with Douglas-fir or Western Larch. Special solid-core construction. Available with MDO or HDO face. Ideal for boat hull construction.	MARINE ¥ A-A ¥ EXT APA ¥ 000 ¥ PS 1-95	A or B	A or B	B	1/4, 3/8, 1/2, 5/8, 3/4	A face & back use S-1 B face or back use S-2	Group 1	Table 12.3 (PDS Table 2) (sanded)
APA APPEARANCE GRADES EXT	Generally applied where a high quality surface is required. Includes APA A-A, A-B, A-C, B-B, B-C, HDO and MDO EXT.(5)	**APA** THE ENGINEERED WOOD ASSOCIATION A-C GROUP 1 EXTERIOR 000 PS 1-95	B or better	C or better	C	1/4, 11/32, 3/8, 15/32, 1/2, 19/32, 5/8, 23/32 3/4	A or C face & back use S-1[6] B face or back use S-2	As Specified	Table 12.3 (PDS Table 1) (sanded)

(1) When exterior glue is specified, i.e. Exposure 1, stress level 2 (S-2) should be used.

(2) Check local suppliers for availability before specifying Plyform Class II grade, as it is rarely manufactured.

(3) Properties and stresses apply only to APA RATED STURD-I-FLOOR and APA RATED SHEATHING manufactured entirely with veneers.

(4) APA RATED STURD-I-FLOOR 2-4-1 may be produced unsanded.

(5) May be available as Structural I. For such designation use Group 1 stresses and Table 2 section properties.

(6) C face and back must be natural unrepaired; if repaired, use stress level 2 (S-2).

Used with permission of APA, the Engineered Wood Association, Tacoma, WA.

Table 12.5 Allowable Stresses for Plywood (APA PDS Table 3)

Allowable Stresses for Plywood (psi) conforming to Voluntary Product Standard PS 1-95 for Construction and Industrial Plywood. Stresses are based on normal duration of load, and on common structural applications where panels are 24 in or greater in width. For other use conditions, see Sec. 3.3 for modifications.

type of stress		species group of face ply	grade stress level[a]				
			S-1		S-2		S-3
			wet	dry	wet	dry	dry only
extreme fiber stress in bending (F_b) tension in plane of plies (F_b) face grain parallel or perpendicular to span (at 45° to face grain use $1/6F_t$)	F_b and F_t	1	1430	2000	1190	1650	1650
		2, 3	980	1400	820	1200	1200
		4	940	1330	780	1110	1110
compression in plane of plies parallel or perpendicular to face grain (at 45° to face grain use $1/3F_c$)	F_c	1	970	1640	900	1540	1540
		2	730	1200	680	1100	1100
		3	610	1060	580	990	990
		4	610	1000	580	950	950
shear through the thickness[c] parallel or perpendicular to face grain (at 45° to face grain use $2F_v$)	F_v	1	155	190	155	190	160
		2, 3	120	140	120	140	120
		4	110	130	110	130	115
rolling shear (in the plane of plies) parallel or perpendicular to face grain (at 45° to face grain use $1^1/3F_s$)	F_s	marine and structural I	63	75	63	75	–
		all other[b]	44	53	44	53	48
modulus of rigidity (or shear modulus) shear in plane perpendicular to plies (through the thickness) (at 45° to face grain use $4G$)	G	1	70,000	90,000	70,000	90,000	82,000
		2	60,000	75,000	60,000	75,000	68,000
		3	50,000	60,000	50,000	60,000	55,000
		4	45,000	50,000	45,000	50,000	45,000
bearing (on face) perpendicular to plane of plies	$F_{c\perp}$	1	210	340	210	340	340
		2, 3	135	210	135	210	210
		4	105	160	105	160	160
modulus of elasticity in bending in plane of plies face grain parallel or perpendicular to span	E	1	1,500,000	1,800,000	1,500,000	1,800,000	1,800,000
		2	1,300,000	1,500,000	1,300,000	1,500,000	1,500,000
		3	1,100,000	1,200,000	1,100,000	1,200,000	1,200,000
		4	900,000	1,000,000	900,000	1,000,000	1,000,000

[a]See Table 12.4a and Table 12.4b [APA PDS pages 12 and 13] for Guide.
To qualify for stress level S-1, gluelines must be exterior and only veneer grades N, A, and C (natural, not repaired) are allowed in either face or back.
For stress level S-2, gluelines must be exterior and veneer grade B, C-plugged and D are allowed on the face or back.
Stress level S-3 includes all panels with interior or intermediate (IMG) gluelines.

[b]Reduce stresses 25% for three-layer (four- or five-ply) panels over 5/8 in thick. Such layups are possible under PS 1-95 for APA rated sheathing, APA rated sturd-I-floor, underlayment, C-C plugged and C-D plugged grades over 5/8 in through 3/4 in thick.

[c]Shear-through-the-thickness stresses for marine and special exterior grades may be increased 33%. See Sec. 3.8.1 for conditions under which stresses for other grades may be increased.

Used with permission of APA, the Engineered Wood Association, Tacoma, WA.

Determine the plywood section properties and allowable stresses for plywood.

Solution:

Table 12.4a — Table 12.4a indicates that unsanded section properties from Table 12.3 should be used in conjunction with the grade stress level S-3. The species group is referred to the key to span rating in Table 12.2. The key to span rating indicates that a 24/0 span rating is available for the $^3/_8$ in thickness, and the grade stress level for species group 1 in Table 12.5 is to be used for allowable plywood stresses.

For a $^3/_8$ in APA rated sheathing, exposure 1 and 24/0 span rating, the following values for stresses applied parallel to face grain (across the roof joist supports) are obtained.

A. Section Properties

Table 12.3 (PDS Table 1) — From Table 12.3, unsanded $^3/_8$ in plywood, $^3/_8$-U; $I = 0.039$ in^4/ft; KS (effective section modulus) $= 0.152$ in^3/ft; Ib/Q (rolling shear constant) $= 3.110$ in^2/ft; and the approximate weight is 1.1 lbf/ft^2.

B. Allowable Stresses

Table 12.5 (PDS Table 3) — From Table 12.5, group 1 stresses for S-3 grade stress level for the dry only condition are: $F_b = 1650$ lbf/in^2; F_s (rolling shear) $= 48$ lbf/in^2; and $E = 1.8 \times 10^6$ lbf/in^2.

Example 12.2
Plywood Shelf Design

A 40 in $\times$ 30 in shelf is made of $^3/_4$ in APA rated sheathing, 40/20 exposure 1. The face grain is across supports spanning 40 in.

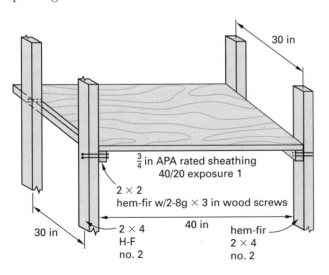

30 in

$\frac{3}{4}$ in APA rated sheathing
40/20 exposure 1

2 × 2
hem-fir w/2-8g × 3 in wood screws

30 in

2 × 4
H-F
no. 2

40 in

hem-fir
2 × 4
no. 2

Determine the allowable uniform load for the shelf considering only the 40 in span direction (the strong direction). Assume a load duration of two months, normal temperature conditions, and dry-use conditions. The allowable deflection is $L/360$.

$^3/_4$ in APA rated sheathing 40/20 exposure 1 is selected. Assume $C_D = 1.15$, $C_M = 1.0$.

Solution:

Table 12.4a — Table 12.4a shows a grade stress level of S-3; the species group = key to span rating; and section property = Table 12.3 (unsanded).

Table 12.2 — The key to span rating for 40/20 shows the species group 4. The section properties for

Table 12.3 — the $^3/_4$ in unsanded panel are given as follows.

$$I = 0.234 \text{ in}^4/\text{ft}$$
$$KS = 0.496 \text{ in}^3/\text{ft}$$
$$Ib/Q = 6.455 \text{ in}^2/\text{ft}$$

Table 12.5 — The allowable stresses for S-3 grade level with the dry only use condition and species group 4 are

$$F_b = 1110 \text{ lbf/in}^2$$
$$F_s = 48 \text{ lbf/in}^2$$
$$E = 1.0 \times 10^6 \text{ lbf/in}^2$$

A. Bending (per foot width of panel) is

$$f_b = \frac{M}{KS} = \frac{\dfrac{wL^2}{(8 \text{ ft})\left(12 \dfrac{\text{in}}{\text{ft}}\right)}}{KS}$$
$$= \frac{wL^2}{(96 \text{ in})(KS)}$$

Table 2.3.2
$$F_b' = F_b C_D C_M C_t = \left(1110 \frac{\text{lbf}}{\text{in}^2}\right)((1.15)(1.0))$$
$$= 1276.5 \text{ lbf/in}^2$$

$w = \text{lbf/ft}$
$L = \text{in}$
$f_b = \text{lbf/in}^2$
$f_s = \text{lbf/in}^2$

Set $f_b = F_b'$ and solve for w.

$$w = \frac{F_b'(96 \text{ in})(KS)}{L^2}$$
$$= \frac{\left(1276.5 \dfrac{\text{lbf}}{\text{in}^2}\right)(96 \text{ in})\left(0.496 \dfrac{\text{in}^3}{\text{ft}}\right)}{(40 \text{ in} + 1.5 \text{ in})^2}$$
$$= 35 \text{ lbf/ft uniform load per foot width of panel}$$

B. Rolling shear (per foot width of panel) is
$$V_{\max} = \frac{wL}{2} = \frac{wL}{24 \text{ in}}$$

$$f_s = \frac{V_{max}}{\frac{Ib}{Q}} = \frac{\frac{wL}{24 \text{ in}}}{6.455 \frac{\text{in}^2}{\text{ft}}} = \frac{wL}{154.92 \frac{\text{in}^3}{\text{ft}}}$$

$$F_s' = F_s C_D C_M = \left(48 \frac{\text{lbf}}{\text{in}^2}\right)((1.15)(1.0))$$

$$= 55.2 \text{ lbf/in}^2$$

Set $f_s = F_s'$ and solve for w.

$$w = \frac{F_s'\left(154.92 \frac{\text{in}^3}{\text{ft}}\right)}{L}$$

$$= \frac{\left(55.2 \frac{\text{lbf}}{\text{in}^2}\right)\left(154.92 \frac{\text{in}^3}{\text{ft}}\right)}{40 \text{ in} + 1.5 \text{ in}}$$

$$= 206 \text{ lbf/ft uniform load per foot}$$
width of panel

C. Deflection (per foot width of panel) is

$$\Delta = \frac{5wL^4}{384EI} \quad [\text{where } w = \text{lbf/ft}]$$

$$= \frac{5wL^4}{\left(4608 \frac{\text{in}}{\text{ft}}\right)EI}$$

$$\delta_{max} = \frac{L}{360}$$

Set $\delta_{max} = \Delta$ and solve for w.

$$w = \frac{(4608 \text{ in})EI}{(360)(5)L^3}$$

$$= \frac{(4608 \text{ in})\left(1.0 \times 10^6 \frac{\text{lbf}}{\text{in}^2}\right)(0.234 \text{ in}^4)}{(360)(5)(40 \text{ in} + 1.5 \text{ in})^3}$$

$$= 8 \text{ lbf/ft uniform load per width}$$
of panel

Therefore, the controlling allowable uniform load is 8 lbf/ft^2.

4. Diaphragms and Shear Walls

Wind and earthquakes are the principal lateral forces a structure must resist. With relatively minor modifications, building walls can act as shear walls to efficiently resist these lateral loads. The floors and roof can also act as diaphragms to redistribute or gather forces and transmit them to the shear walls.

A. Diaphragms

A *blocked diaphragm* is a diaphragm in which all panel edges occur over and are fastened to common framing lumber. An *unblocked diaphragm* is a diaphragm in which only panel edges in one direction occur over and are fastened to common framing lumber (as in a typical roof diaphragm for standard residential construction).

Floors and roofs can be designed to act as very deep, horizontal beams that carry the lateral forces applied to the walls between the floors and roof. These deep beams, or diaphragms, span between shear walls and other structural elements carrying the lateral loads to the building foundation.

In conventional wood frame buildings, the lateral forces due to wind or seismic forces are primarily carried by the wall frames. In other cases, however, the lateral forces are carried by the wall framing to the horizontal diaphragms at the top and then transferred to the foundation through the vertical wall elements (shear walls). These shear walls used with plywood horizontal diaphragms (roofs and floors) can also be masonry or concrete.

The design of diaphragm panels subjected to the loads normal to the surface of the sheathing must be completed prior to the investigation of diaphragm actions. The diaphragm is a flat structural member acting like a deep beam, and it acts as a web for resisting shear, while the diaphragm edge members perform the function of flanges for resisting bending stresses (Fig. 12.4). The edge members are called "chords" in the diaphragm system, and they may consist of dimension lumber (top plates), joists, studs, trusses, and so on. Due to the large depth of diaphragms in the direction parallel to the applications of lateral loads, shear forces are considered uniform across the depth of the diaphragm (Fig. 12.4(b)). The diaphragm chords carry all flange forces, acting in compression and tension, to resist bending stresses in the diaphragm system.

The horizontal (in-plane) load-carrying capacity of the diaphragm depends on the nail strength and also on whether the diaphragms are blocked or unblocked. Blocking usually consists of 2×4s or 2×6s fastened between the joists, roof rafters, or other primary structural members, and support frames for connecting the edges of the plywood panels. The purpose of blocking is to allow increased shear transfer. Buckling of unsupported panel edges controls unblocked panel loads. An unblocked diaphragm is one having two of its panel's four edges not supported by lumber framing.

It is important to note that a diaphragm system consists of three parts: the web (diaphragm panels), the chords (edge members), and the connections that hold the web and the chords as well as transfer loads to the other structural elements and to the foundation.

Just as with other beams, diaphragms are designed to resist the imposed shear and bending stresses. The shear stress is carried by the plywood decking and is assumed to be uniformly distributed across the depth. The nailing schedule and panel splicing details required for a given design shear force are determined from Table 12.6.

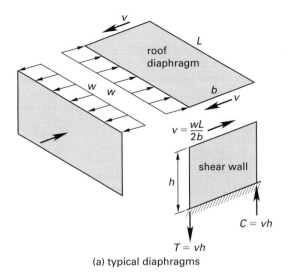

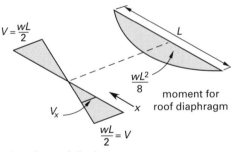

(b) diaphragm chords

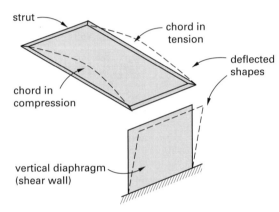

shear for roof diaphragm

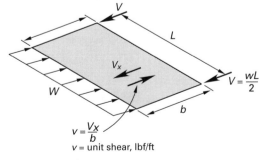

(c) diaphragm shear and moments

Figure 12.4 Plywood Sheathing Panels for Diaphragms

In both Tables 12.6 and 12.7, blocking the unsupported plywood edges between the principal framing members significantly increases the shear capacity. Blocking means installing short pieces of lumber to which all abutting free edges are attached.

The bending forces in diaphragms are usually resisted by the roof or floor perimeter framing, which acts as the diaphragm chords—comparable to the chords of a truss. The chord is sized to resist the calculated axial and bending forces. Since diaphragms are commonly much longer than available or manageable lumber lengths, the chords must be spliced adequately.

Example 12.3
Horizontal Roof Diaphragm

A wood frame building is shown in the following illustration. Design data are $L = 80$ ft, $b = 38$ ft, $h = 16$ ft (mean roof height), wind pressure on vertical surface $= 33.25$ lbf/ft², framing members $= 2 \times 6$ lumbers, roof diaphragm selected (based on the load perpendicular to the plywood panel surface) $= {}^{15}\!/_{32}$ in, and APA rated sheathing with 8d common wire nails.

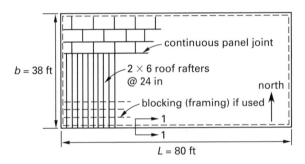

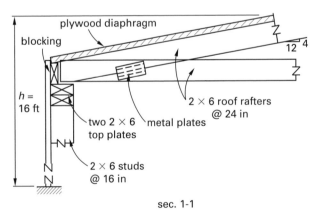

sec. 1-1

Determine the diaphragm requirements.

Solution:

A. Wind Pressure on South Wall

The lateral force is perpendicular to the continuous panel joint and perpendicular to the unblocked panel edge. Therefore, this is an example of case 1 (see Table 12.6).

Table 12.6 Required Panel Details: Horizontal Diaphragms (APA Design/Construction Guide: Diaphragms and Shear Walls Table 1)

RECOMMENDED SHEAR (POUNDS PER FOOT) FOR HORIZONTAL APA PANEL DIAPHRAGMS WITH FRAMING OF DOUGLAS-FIR, LARCH OR SOUTHERN PINE[a] FOR WIND OR SEISMIC LOADING

Panel Grade	Common Nail Size	Minimum Nail Penetration in Framing (inches)	Minimum Nominal Panel Thickness (inch)	Minimum Nominal Width of Framing Member (inches)	Blocked Diaphragms Nail Spacing (in.) at diaphragm boundaries (all cases), at continuous panel edges parallel to load (Cases 3 & 4), and at all panel edges (Cases 5 & 6)[b]				Unblocked Diaphragms Nails Spaced 6" max. at Supported Edges[b]	
					6	4	2-1/2[c]	2[c]	Case 1 (No unblocked edges or continuous joints parallel o load)	All other configurations (Cases 2, 3, 4, 5 & 6)
					Nail Spacing (in.) at other panel edges (Cases 1, 2, 3 & 4)[b] †					
					6	6	4	3		
APA STRUCTURAL I grades	6d[e]	1-1/4	5/16	2	185	250	375	420	165	125
				3	210	280	420	475	185	140
	8d	1-3/8	3/8	2	270	360	530	600	240	180
				3	300	400	600	675	265	200
	10d[d]	1-1/2	15/32	2	320	425	640	730	285	215
				3	360	480	720	820	320	240
APA RATED SHEATHING APA RATED STURD-I-FLOOR and other APA grades except Species Group 5	6d[e]	1-1/4	5/16	2	170	225	335	380	150	110
				3	190	250	380	430	170	125
			3/8	2	185	250	375	420	165	125
				3	210	280	420	475	185	140
	8d	1-3/8	3/8	2	240	320	480	545	215	160
				3	270	360	540	610	240	180
			7/16	2	255	340	505	575	230	170
				3	285	380	570	645	255	190
			15/32	2	270	360	530	600	240	180
				3	300	400	600	675	265	200
	10d[d]	1-1/2	15/32	2	290	385	575	655	255	190
				3	325	430	650	735	290	215
			19/32	2	320	425	640	730	285	215
				3	360	480	720	820	320	240

(a) For framing of other species: (1) Find specific gravity for species of lumber in the AFPA National Design Specification. (2) Find shear value from table above for nail size for actual grade. (3) Multiply value by the following adjustment factor: Specific Gravity Adjustment Factor = [1 – (0.5 – SG)], where SG = specific gravity of the framing. This adjustment shall not be greater than 1.

(b) Space nails maximum 12 inches o.c. along intermediate framing members (6 in. o.c. when supports are spaced 48 in. o.c. or greater). Fasteners shall be located 3/8 inch from panel edges.

(c) Framing at adjoining panel edges shall be 3-in. nominal or wider, and nails shall be staggered where nails are spaced 2 inches o.c. or 2-1/2 inches o.c.

(d) Framing at adjoining panel edges shall be 3-in. nominal or wider, and nails shall be staggered where 10d nails having penetration into framing of more than 1-5/8 inches are spaced 3 inches o.c.

(e) 8d is recommended minimum for roofs due to negative pressures of high winds.

Notes: Design for diaphragm stresses depends on direction of continuous panel joints with reference to load, not on direction of long dimension or strength axis of sheet. Continuous framing may be in either direction for blocked diaphragms.

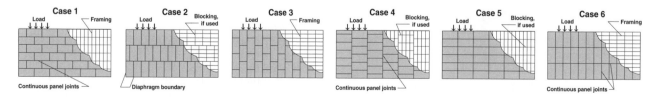

Used with permission of APA, the Engineered Wood Association, Tacoma, WA.

The horizontal load acting on the diaphragm is

$$w = (\text{wind pressure})\left(\frac{\text{wall height}}{2}\right)$$

$$= \left(33.25 \; \frac{\text{lbf}}{\text{ft}^2}\right)\left(\frac{16 \; \text{ft}}{2}\right) = 266 \; \text{lbf/ft}$$

The shear at the east and west walls is

$$V = \frac{wL}{2} = \frac{\left(266 \; \dfrac{\text{lbf}}{\text{ft}}\right)(80 \; \text{ft})}{2} = 10{,}640 \; \text{lbf}$$

The unit shear stress along the east and west walls is

$$v = \frac{V}{b} = \frac{10{,}640 \; \text{lbf}}{38 \; \text{ft}} = 280 \; \text{lbf/ft}$$

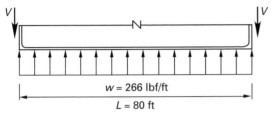

(a) horizontal load on diaphragm

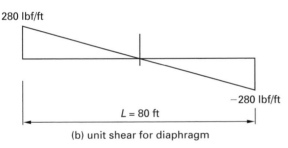

(b) unit shear for diaphragm

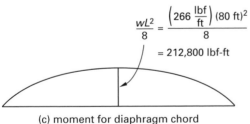

(c) moment for diaphragm chord

The allowable unit shears from Table 12.6 are as follows.

For unblocked diaphragms,

$$v = 240 \; \text{lbf/ft} < 280 \; \text{lbf/ft} \quad [\text{no good}]$$

For blocked diaphragms,

$$v = 360 \; \text{lbf/ft} > 280 \; \text{lbf/ft} \quad [\text{OK}]$$

Thus, along the east or west wall, use the blocked diaphragm with nail spacings at 4 in along the diaphragm

boundary, 6 in along continuous panel joints as well as along other plywood edges, and 12 in along intermediate framing members.

If reduced unit shears toward the center of the diaphragm (away from east or west wall) are taken into consideration, the nail spacings can be increased and/or blocking can be omitted.

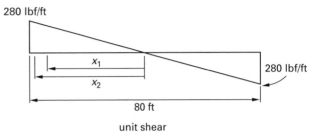

unit shear

For a nail spacing of 6 in at the diaphragm boundary and at other panel edges (see Table 12.6), the allowable shear for unblocked diaphragm = 240 lbf/ft, and allowable shear for blocked diaphragm = 270 lbf/ft.

The blocking along the continuous panel joints can be omitted at a distance from the center of the diaphragm (with nails spaced 6 in at supported edges) of

$$x_1 = \frac{240 \; \dfrac{\text{lbf}}{\text{ft}}}{\dfrac{280 \; \dfrac{\text{lbf}}{\text{ft}}}{40 \; \text{ft}}} = 34.3 \; \text{ft}$$

Six in nail spacings at the diaphragm boundary for which blocking is provided can begin at a distance from the diaphragm center as follows.

$$x_2 = \frac{270 \; \dfrac{\text{lbf}}{\text{ft}}}{\dfrac{280 \; \dfrac{\text{lbf}}{\text{ft}}}{40 \; \text{ft}}} = 38.6 \; \text{ft}$$

The axial forces in the diaphragm chord (80 ft in length) are determined from the maximum diaphragm moment, which is resolved into a couple. The maximum moment is

$$M_{\max} = \frac{wL^2}{8} = \frac{\left(266 \; \dfrac{\text{lbf}}{\text{ft}}\right)(80 \; \text{ft})^2}{8}$$

$$= 212{,}800 \; \text{lbf-ft}$$

The tension or compression force in the diaphragm is

$$\frac{M_{\max}}{b} = \frac{212{,}800 \; \text{lbf-ft}}{38 \; \text{ft}}$$

$$= 5600 \; \text{lbf}$$

The chord member can be top plates at the top of the stud wall, masonry wall with reinforcement, and so on, and they will be designed for 5600 lbf tension and compression forces.

B. Wind Pressure on East Wall

The lateral force is parallel to the continuous panel joint and parallel to the unblocked panel edge. Therefore, this is case 3 (see Table 12.6).

The horizontal load acting on the diaphragm is

$$w = \left(33.25 \; \frac{\text{lbf}}{\text{ft}^2}\right)\left(\frac{16 \; \text{ft}}{2}\right) = 266 \; \text{lbf/ft}$$

The shear at the east and west walls is

$$V = \frac{wb}{2} = \frac{\left(266 \; \dfrac{\text{lbf}}{\text{ft}}\right)(38 \; \text{ft})}{2} = 5054 \; \text{lbf}$$

The unit shear stress along north and south walls is

$$v = \frac{V}{L} = \frac{5054 \; \text{lbf}}{80 \; \text{ft}} = 63.2 \; \text{lbf/ft}$$

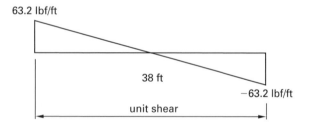

The allowable unit shears from Table 12.6 are as follows.

For unblocked diaphragms with nails spaced 6 in at supported edge, $v = 180$ lbf/ft.

For blocked diaphragms with nails spaced 4 in at boundaries and spaced 6 in at other panel edges, $v = 360$ lbf/ft.

Since the allowable unit shear for unblocked diaphragms, 180 lbf/ft, is greater than the required unit shear, 63.2 lbf/ft, unblocked diaphragms are allowed as seen in Case 3 in Table 12.6.

B. Shear Walls

Building walls that are parallel to an applied lateral force can carry that force down to the foundation as do short, deep cantilevers. Once the shear forces along the shear wall-diaphragm intersection are determined, Table 12.7 can be used to determine the plywood thickness, panel layout, and nailing schedule required to provide the design capacity.

C. Design Methods: Shear Walls and Diaphragms

The basic design procedure is to determine the applied loads and detail the respective elements to carry the loads.

step 1:   Calculate the applied loads as shears (lbf/ft) along the supported edge of diaphragms or the loaded edge of shear walls.

step 2:   Determine panel layout, plywood thickness, and nailing schedule from Table 12.6 or 12.7.

step 3:   Determine diaphragm chord size and detail splices.

step 4:   Check deflections by comparing length-width ratios to allowable ones.

step 5:   Detail connection between elements and to the foundation.

Example 12.4
Diaphragm and Shear Wall

Given the following building, determine the design shear on the diaphragms and shear walls of the building shown. Design and detail the roof as a diaphragm and the first-floor interior wall as a shear wall. Assume a wind load of 25 lbf/ft² and consider only wind against the long side of the building. Use southern pine no. 3 grade for the framing.

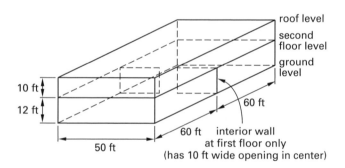

Solution:

A. Determine the wind loads applied to the long edges of the roof and floor from the wall that frames between them. Assume a tributary area distribution of the uniform load.

The wind load on the 120 ft length of the roof is

$$\begin{aligned}
w_{\text{roof}} &= w_{\text{ft}^2}(\text{tributary height}) \\
&= \left(25 \; \frac{\text{lbf}}{\text{ft}^2}\right)\left(\frac{10 \; \text{ft}}{2}\right) \\
&= 125 \; \text{lbf/ft}
\end{aligned}$$

The wind load on the 120 ft length of the second floor is

$$\begin{aligned}
w_{\text{second floor}} &= w_{\text{ft}^2}(\text{tributary height}) \\
&= \left(25 \; \frac{\text{lbf}}{\text{ft}^2}\right)\left(\frac{10 \; \text{ft} + 12 \; \text{ft}}{2}\right) \\
&= 275 \; \text{lbf/ft}
\end{aligned}$$

Table 12.7 APA Panel Shear Wall Capacities (APA Design/Construction Guide: Diaphragms Table 2)

RECOMMENDED SHEAR (POUNDS PER FOOT) FOR APA PANEL SHEAR WALLS WITH FRAMING OF DOUGLAS-FIR, LARCH, OR SOUTHERN PINE[a] FOR WIND OR SEISMIC LOADING[b]

Panel Grade	Minimum Nominal Panel Thickness (in.)	Minimum Nail Penetration in Framing (in.)	Panels Applied Direct to Framing					Panels Applied Over 1/2" or 5/8" Gypsum Sheathing				
			Nail Size (common or galvanized box)	Nail Spacing at Panel Edges (in.)				Nail Size (common or galvanized box)	Nail Spacing at Panel Edges (in.)			
				6	4	3	2[e]		6	4	3	2[e]
APA STRUCTURAL I grades	5/16	1-1/4	6d	200	300	390	510	8d	200	300	390	510
	3/8	1-3/8	8d	230[d]	360[d]	460[d]	610[d]	10d	280	430	550[f]	730
	7/16			255[d]	395[d]	505[d]	670[d]					
	15/32			280	430	550	730					
	15/32	1-1/2	10d	340	510	665[f]	870		—	—	—	—
APA RATED SHEATHING; APA RATED SIDING[g] and other APA grades except species Group 5	5/16 or 1/4[c]	1-1/4	6d	180	270	350	450	8d	180	270	350	450
	3/8			200	300	390	510		200	300	390	510
	3/8	1-3/8	8d	220[d]	320[d]	410[d]	530[d]	10d	260	380	490[f]	640
	7/16			240[d]	350[d]	450[d]	585[d]					
	15/32			260	380	490	640					
	15/32	1-1/2	10d	310	460	600[f]	770		—	—	—	—
	19/32			340	510	665[f]	870		—	—	—	—
APA RATED SIDING[g] and other APA grades except species Group 5			Nail Size (galvanized casing)					Nail Size (galvanized casing)				
	5/16[c]	1-1/4	6d	140	210	275	360	8d	140	210	275	360
	3/8	1-3/8	8d	160	240	310	410	10d	160	240	310[f]	410

(a) For framing of other species: (1) Find specific gravity for species of lumber in the AFPA National Design Specification. (2) For common or galvanized box nails, find shear value from table above for nail size for actual grade. (3) Multiply value by the following adjustment factor: Specific Gravity Adjustment Factor = [1 − (0.5 − SG)], where SG = specific gravity of the framing. This adjustment shall not be greater than 1.

(b) All panel edges backed with 2-inch nominal or wider framing. Install panels either horizontally or vertically. Space nails maximum 6 inches o.c. along intermediate framing members for 3/8-inch and 7/16-inch panels installed on studs spaced 24 inches o.c. For other conditions and panel thicknesses, space nails maximum 12 inches o.c. on intermediate supports. Fasteners shall be located 3/8 inch from panel edges.

(c) 3/8-inch or APA RATED SIDING 16 oc is minimum recommended when applied direct to framing as exterior siding.

(d) Shears may be increased to values shown for 15/32-inch sheathing with same nailing provided (1) studs are spaced a maximum of 16 inches o.c., or (2) if panels are applied with strength axis across studs.

(e) Framing at adjoining panel edges shall be 3-inch nominal or wider, and nails shall be staggered where nails are spaced 2 inches o.c. Check local code for variations of these requirements.

(f) Framing at adjoining panel edges shall be 3-inch nominal or wider, and nails shall be staggered where 10d nails having penetration into framing of more than 1-1/2 inches are spaced 3 inches o.c. Check local code for variations of these requirements.

(g) Values apply to all-veneer plywood APA RATED SIDING panels only. Other APA RATED SIDING panels may also qualify on a proprietary basis. APA RATED SIDING 16 oc plywood may be 11/32 inch, 3/8 inch or thicker. Thickness at point of nailing on panel edges governs shear values.

TYPICAL LAYOUTS FOR SHEAR WALLS

Load Framing

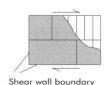

Shear wall boundary

Blocking

Framing

Foundation resistance

Used with permission of APA, the engineered Wood Association, Taxoma, WA.

Note that the wind load on the lower 6 ft of the wall is carried directly to the foundation and represents a relatively small shear force.

B. Determine the shears along the supported edges of the diaphragms (where they are connected to the shear walls). The end walls act as shear walls.

The roof-to-end walls shear is

$$V_{\text{roof-to-end wall}} = \frac{w_{\text{roof}}\left(\frac{1}{2}L\right)}{\text{end wall length}}$$
$$= \frac{\left(125 \, \frac{\text{lbf}}{\text{ft}}\right)\left(\frac{120 \text{ ft}}{2}\right)}{50 \text{ ft}}$$
$$= 150 \text{ lbf/ft}$$

The floor-to-end walls shear is

$$V_{\text{floor-to-end walls}} = \frac{w_{\text{second floor}}\left(\frac{3}{8}\right)\left(\frac{1}{2}L\right)}{\text{end wall length}}$$
$$= \frac{\left(275 \, \frac{\text{lbf}}{\text{ft}}\right)\left(\frac{3}{8}\right)(60 \text{ ft})}{50 \text{ ft}}$$
$$= 124 \text{ lbf/ft}$$

The floor-to-centerline walls shear is

$$V_{\text{floor to centerline}} = \frac{w_{\text{second floor}}\left(\frac{10}{8}\right)\left(\frac{1}{2}L\right)}{\text{interior wall length}}$$
$$= \frac{\left(275 \, \frac{\text{lbf}}{\text{ft}}\right)\left(\frac{10}{8}\right)(60 \text{ ft})}{40 \text{ ft}}$$
$$= 516 \text{ lbf/ft}$$

Note that the 10 ft opening is considered by loading only 40 ft of the interior wall.

C. To design the roof as a horizontal diaphragm, first find an adequate configuration in Table 12.6. Because installing complete blocking is expensive, first try to find an option with unblocked edges. The lower group of panel grades is less expensive, so also try to find a layout in that section of Table 12.6. Several layouts meet the design requirement of 150 lbf/ft. In practice, the choice might be influenced by snow loads, local practice, or other architectural considerations. For this case, assume 2 in nominal framing will be adequate and that $^3/_8$ in panel will be adequate for the wind load. A 160 lbf/ft capacity (requirement is 150 lbf/ft) is provided by the following configuration: APA rated sheathing, $^3/_8$ in panel thickness, 8d nails on 6 in centers along the supported edge (i.e., end wall) and on 12 in centers along the other framing. The edges can be unblocked,

and 2 in nominal framing members are adequate. The panel layout can be case 2, 3, 4, 5, or 6.

The chord force is determined by finding the maximum bending moment and dividing by the diaphragm depth—the moment arm of the chord forces.

The maximum moment is

$$M_{\text{max}} = \frac{wL^2}{8}$$
$$= \frac{\left(125 \, \frac{\text{lbf}}{\text{ft}}\right)(120 \text{ ft})^2}{8}$$
$$= 225{,}000 \text{ ft-lbf}$$

Therefore, the bending force in the chords is

$$\frac{M_{\text{max}}}{d} = \frac{225{,}000 \text{ ft-lbf}}{50 \text{ ft}}$$
$$= 4500 \text{ lbf}$$

Since one chord will be in tension and the other in compression, the smaller of the two allowable axial stresses will control the chord size. NDS Supplement Table 4B gives $F_c = 925 \text{ lbf/in}^2$ and $F_t = 425 \text{ lbf/in}^2$ for southern pine no. 3 grade, so the tension chord controls the design size for an assumed 6 in deep member. The required chord area is, therefore,

$$A_{\text{chord}} = \frac{\dfrac{M_{\text{max}}}{d}}{F_t(\text{load duration factor})}$$
$$= \frac{4500 \text{ lbf}}{\left(425 \, \dfrac{\text{lbf}}{\text{in}^2}\right)(1.60)}$$
$$= 6.62 \text{ in}^2$$

1.60 is the load duration factor for wind loads (NDS Table 2.3.2). Assuming a doubled 2 in nominal chord, splices will have a 1.5 in wide continuous member. The required depth of the chord is

$$d_{\text{chord}} = \frac{A_{\text{chord}}}{\text{chord width}}$$
$$= \frac{6.62 \text{ in}^2}{1.5 \text{ in}}$$
$$= 4.41 \text{ in}$$

Even after removing some of the chord section for the splice fasteners, any chord of doubled 2 × 6s will be adequate. The splices should be designed to handle the maximum chord force.

The diaphragm is only as good as the connection supporting it at the shear walls. This connection should have a design capacity of 160 lbf/ft, the capacity of the diaphragm—not the 150 lbf/ft design load. There are several ways to achieve this capacity. One efficient way is to have the panel wall sheathing overlap the roof perimeter framing.

The Uniform Building Code (and most other codes) specifies a maximum length:width ratio of 4:1 for horizontal diaphragms sheathed with plywood. This ratio is intended to eliminate excessive deflection. The ratio in this example is 120 ft:50 ft = 2.4:1, so the criterion is satisfied.

D. The interior wall has a shear load of 516 lbf/ft, applied on the top by the second floor and transmitted down to the foundation. Table 12.7 distinguishes between walls with the plywood applied directly to the framing and plywood applied over $^{1}/_{2}$ in gypsum sheathing intended as fire walls. Assuming this is not a fire wall, a 530 lbf/ft load (requirement is 516 lbf/ft) can be carried in a shear wall with the following specifications: APA rated sheathing, $^{3}/_{8}$ in panel thickness, with 8d nails 2 in on center at the boundary framing members and 12 in on center at all other framing. Any of the panel layouts illustrated at the bottom of Table 12.7 are acceptable.

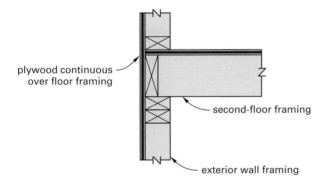

plywood continuous over floor framing

second-floor framing

exterior wall framing

The chords of shear walls are the vertical framing members at the corners and around openings. Since standard framing practice results in triple members at corners and double members at openings, the chords at the openings control. The load at the top of each half of the shear wall is

(shear at diaphragm edge)(diaphragm length)

$$= \left(516 \ \frac{\text{lbf}}{\text{ft}}\right)(20 \text{ ft}) = 10{,}320 \text{ lbf}$$

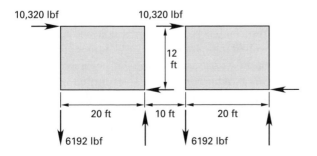

10,320 lbf   10,320 lbf

12 ft

20 ft   10 ft   20 ft

6192 lbf   6192 lbf

This force creates a bending moment that is maximum at the bottom of the wall, where it is equal to

$$M_{\text{bending}} = w_{\text{top of shear wall}}h$$
$$= (10{,}320 \text{ lbf})(12 \text{ ft})$$
$$= 123{,}840 \text{ ft-lbf}$$

The height of the wall is 12 ft. This causes a bending force in the vertical chords equal to the applied moment divided by the chord separation, or shear wall length. The chord force is

$$V_{\text{chord}} = \frac{M_{\text{bending}}}{L_{\text{shear wall}}}$$
$$= \frac{123{,}840 \text{ ft-lbf}}{20 \text{ ft}}$$
$$= 6192 \text{ lbf}$$

With the allowable tension stress calculated for the roof diaphragm, this chord force requires an area of

$$A_{\text{chord force}} = \frac{6192 \text{ lbf}}{C_D F_t}$$
$$= \frac{6192 \text{ lbf}}{(1.60)\left(425 \ \dfrac{\text{lbf}}{\text{in}^2}\right)}$$
$$= 9.1 \text{ in}^2$$

Assuming the verticals will be short enough to be one piece, the full 3 in width will be available. This requires a vertical framing member depth of

$$D_{\text{member}} = \frac{A_{\text{chord force}}}{w_{\text{vertical members}}}$$
$$= \frac{9.1 \text{ in}^2}{3 \text{ in}}$$
$$= 3.03 \text{ in}$$

This is less than the 3.5 in of a 2 × 4, so a 2 × 4 wall is adequate.

The 2 × 4 wall must be connected to the foundation to resist both the shear load of 530 lbf/ft (requirement is 516 lbf/ft) and the uplift load in the chords of 6192 lbf at each corner and side of the opening. Anchor bolts at closer than standard spacings or special steel fittings can be used to provide this connection capacity.

The height:width ratio for this shear wall is 12 ft:40 ft, or 0.30:1, less than the 1:1 ratio for which deflections should be calculated. Both 20 ft sections of the interior wall are taken into account. This issue of openings in plywood diaphragms and shear walls is very complex and is at the cutting edge of the research in the field. This text has necessarily treated this subject lightly.

5. Plywood-Lumber Built-Up Beams

[PDS Supplement #2:
Design and Fabrication of Plywood-Lumber Beams, 1992]

With its alternating grain directions, plywood is very strong in shear. Its panel configuration also makes plywood convenient to use for webs of built-up beams. Lumber, with all its fibers oriented in the same direction, can efficiently resist the axial forces found in the flanges of built-up beams. For spans longer or loads heavier than dimension lumber beams can handle, plywood-lumber beams can be cheaper and easier to obtain than glued laminated members.

There are many design variables involved in designing a plywood-lumber beam. The beam depth and configuration can be changed. The size, number, species, and grade of the lumber flanges are variable. Finally, the webs can be any number of any thickness and type of plywood.

With all these available design variables, there are two basic design methodologies. Several manufacturers' groups and government agencies publish tables of plywood-lumber beam capacities, spans, and details.

The other choice is to pick a trial section based on experience and available material. The shear stresses, bending stresses, and deflections are then checked. If the trial section is adequate without being unreasonably understressed, the design is completed by detailing the connections.

Trying to achieve an optimal built-up beam design with all the design variables could be a long task. Unless the beams are to be mass-produced, it is generally uneconomical to spend a lot of time trying to optimize design. A more important consideration is material availability.

The following design procedure is based on *PDS Supplement #2: Design and Fabrication of Plywood-Lumber Beams*. This publication contains much valuable information on fabrication detailing, asymmetrical sections, and detailing. Its use is highly recommended.

A. Design Considerations

The two basic configuration layouts used in plywood-lumber beams are the I-beam and the box beam.

The allowable stresses for the lumber flanges are found in the *Supplement, National Design Specification*.

Deflections are calculated through standard mechanics procedures and should be limited to the same values as solid beams. Since these beams tend to be long and/or heavily loaded and are fabricated instead of sawn, camber is a design consideration. The recommended camber is 1.5 times the dead load deflection.

B. Trial Section

To determine a trial section, choose the beam depth as $^1/_8$ to $^1/_{12}$ of the span, with a depth that makes efficient use of the 4 ft module of commercially available

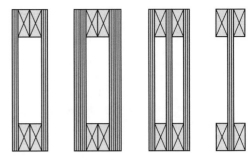

typical section in outer portions

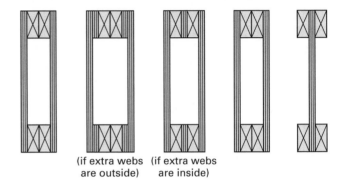

(if extra webs   (if extra webs
are outside)   are inside)

typical section in center portion

Figure 12.5 Plywood-Lumber Beam Cross Sections

Used with permission of APA, the Engineered Wood Association, Tacoma, WA.

plywood. Table 12.8 gives very approximate values for the shear and moment capacities of various cross sections. These are only preliminary cross sections, and they must be checked against the actual design and allowable stresses.

C. Lumber Flanges

Since the lumber flanges are primarily loaded axially by the bending stresses, F_c and F_t are the considered allowable stresses. If the beam is symmetrical, the smaller of the two allowable axial stresses will control. The usual bending stress equation is used with the net moment of inertia, I_n, which is calculated neglecting those longitudinal fibers in the plywood webs and lumber flanges that are interrupted by butt splices.

When calculating I_n, three reductions in the lumber flange area must be considered. The first is due to the resurfacing required for a competent glue joint at the web/flange connection. Assume this is a $^1/_8$ in reduction in width across the beam cross section. The second area reduction is due to resurfacing the depth of the assembled beam and is intended to smooth out irregularities in the fit along the top and bottom of the beam. Assume this reduction in beam and flange depth to be $^3/_8$ in for beams less than 24 in deep and $^1/_2$ in for beams 24 in deep or deeper.

Table 12.8 Preliminary Capacities of Plywood-Lumber Beam Cross Sections (APA PDS Supplement Two, Design and Fabrication of Glued Plywood-Lumber Beams, Appendix B)

Plywood webs, butt joints staggered 24" minimum, spliced per PDS Section 5.6.3.2

Continuous lumber flanges (no butt joints), resurfaced, for gluing per Part 1, Section 4.1.2

*See page 21 for notes.

Depth, Flange $V_{horizontal}$ [5][6]	Max. Moment,[1][2] M (ft-lb) M_{flange}	M_{web} [3][4]	M_{total}	Max. Shear,[1] V_h (lb)
12" 1-2x4	3563	538	4101	1145
2-2x4	7126	538	7664	1159
3-2x4	10690	538	11227	1165
16" 1-2x4	5657	971	6629	1612
2-2x4	11314	971	12286	1648
3-2x4	16971	971	17943	1663
1-2x6	5863	842	6704	1486
2-2x6	11725	842	12567	1497
3-2x6	17588	842	18430	1502
20" 1-2x4	7826	1533	9358	2073
2-2x4	15652	1533	17184	2135
3-2x4	23477	1533	25010	2162
1-2x6	8639	1328	9968	1950
2-2x6	17279	1328	18607	1978
3-2x6	25918	1328	27247	1990
1-2x8	8632	1226	9858	1845
2-2x8	17264	1226	18490	1856
3-2x8	25896	1226	27122	1861
24" 1-2x4	9831	2197	12028	2515
2-2x4	19662	2197	21859	2606
3-2x4	29493	2197	31690	2647
1-2x6	11389	1904	13294	2407
2-2x6	22779	1904	24683	2457
3-2x6	34168	1904	36073	2477
1-2x8	11820	1758	13578	2295
2-2x8	23639	1758	25397	2321
3-2x8	35459	1758	37217	2331
1-2x10	11452	1611	13064	2183
2-2x10	22905	1611	24516	2192
3-2x10	34357	1611	35969	2196
30" 2-2x4	26229	3463	29691	3317
3-2x4	39343	3463	42806	3381
4-2x4	52457	3463	55920	3418
2-2x6	31666	3001	34667	3189
3-2x6	47498	3001	50500	3226
4-2x6	63331	3001	66332	3246
2-2x8	34101	2770	36871	3051
3-2x8	51151	2770	53921	3074
4-2x8	68201	2770	70972	3086
2-2x10	34397	2539	36936	2898
3-2x10	51595	2539	54134	2910
4-2x10	68793	2539	71332	2917
2-2x12	32692	2309	35001	2768
3-2x12	49039	2309	51347	2773
4-2x12	65385	2309	67693	2776

Depth, Flange $V_{horizontal}$ [5][6]	Max. Moment,[1][2] M (ft-lb) M_{flange}	M_{web} [3][4]	M_{total}	Max. Shear,[1] V_h (lb)
36" 2-2x4	32842	5015	37856	4014
3-2x4	49262	5015	54277	4104
4-2x4	65683	5015	70698	4156
2-2x6	40721	4346	45067	3915
3-2x6	61081	4346	65427	3971
4-2x6	81441	4346	85787	4003
2-2x8	44930	4012	48942	3785
3-2x8	67395	4012	71407	3823
4-2x8	89860	4012	93872	3844
2-2x10	46605	3677	50283	3626
3-2x10	69908	3677	73585	3650
4-2x10	93210	3677	96888	3663
2-2x12	45487	3343	48831	3475
3-2x12	68231	3343	71574	3489
4-2x12	90975	3343	94318	3497
42" 2-2x6	49871	5939	55810	4632
3-2x6	74806	5939	80746	4711
4-2x6	99742	5939	105681	4755
2-2x8	55968	5482	61450	4515
3-2x8	83952	5482	89434	4571
4-2x8	111936	5482	117418	4602
2-2x10	59220	5026	64245	4360
3-2x10	88830	5026	93855	4398
4-2x10	118440	5026	123465	4419
2-2x12	58956	4569	63525	4202
3-2x12	88434	4569	93003	4227
4-2x12	117912	4569	122481	4241
48" 2-2x6	59080	7781	66861	5340
3-2x6	88621	7781	96402	5443
4-2x6	118161	7781	125942	5502
2-2x8	67135	7182	74318	5240
3-2x8	100703	7182	107885	5316
4-2x8	134270	7182	141453	5358
2-2x10	72087	6584	78670	5093
3-2x10	108130	6584	114714	5147
4-2x10	144173	6584	150757	5177
2-2x12	72843	5985	78829	4935
3-2x12	109265	5985	115250	4973
4-2x12	145687	5985	151672	4994

(Continued)

Table 12.8 (continued) Preliminary Capacities of Plywood-Lumber Beam Cross Sections (PDS Supplement Two, Design and Fabrication of Glued Plywood-Lumber Beams, Appendix B)

Bases and Adjustments:

(1) Basis: Normal duration of load (C_D): 1.00

Adjustments: 0.90 for permanent load (over 50 years)

       1.15 for 2 months, as for snow

       1.25 for 7 days

       1.6 for 10 minutes, as for wind or earthquake

       2.00 for impact

(2) Basis: F_t of flange = 1,000 psi, corrected by C_F (Douglas fir-larch Select Structural, 1997 NDS)

2×4 = 1,000 x 1.5 = 1,500 psi
2×6 = 1,000 x 1.3 = 1,300 psi
2×8 = 1,000 x 1.2 = 1,200 psi
2×10 = 1,000 x 1.1 = 1,100 psi
2×12 = 1,000 x 1.0 = 1,000 psi

Adjustment: $\dfrac{F_t}{1,000}$ for other tabulated tension stresses.

       (C_F in numerator and denominator cancel when flanges are same width.) Also see PDS Section 5.7.3 for adjustments due to butt joints.

(3) Basis: One web effective in bending because web joints are assumed to be unspliced.

Adjustment: 2.0 for web splices per PDS 5.6.1.

(4) Basis: $A_{\parallel}$ of webs for 15/32" or 1/2" APA RATED SHEATHING EXP 1 (CDX). See PDS Tables 1 and 2.

The last flange area reduction results from butt-splicing the lumber to achieve the required beam length. The reduction is a function of the butt-joint spacings. The butt-jointed flange members are neglected in any I_n calculation. If two flange pieces are butt-jointed at a spacing less than 10 times the thickness of the flange pieces, both flange pieces are neglected at the more critical of the two locations. Unless the splices are spaced more than 50 times the flange member width, the unjointed flange members also have their areas reduced according to the factor from Table 12.9. When checking flange tension stresses, the jointed members are neglected, the unjointed members reduced with Table 12.9, and the allowable stress is further reduced by 20%.

The plywood webs contribute some bending resistance. This is accounted for in the I_n calculation by including only the plywood plies parallel to the longitudinal beam axis. Butt-spliced web members must be neglected in this calculation unless they are spliced full depth with a plywood scab.

Adjustments: 0.81 for 3/8"

       0.97 for 3/8" STRUCTURAL I

       1.19 for 15/32" or 1/2" STRUCTURAL I

       1.02 for 19/32" or 5/8"

       1.51 for 19/32" or 5/8" STRUCTURAL I

       1.42 for 23/32" or 3/4"

       1.84 for 23/32" or 3/4" STRUCTURAL I

(5) Basis: t_s of webs for 15/32" or 1/2" APA RATED SHEATHING EXP 1 (CDX). See PDS Tables 1 and 2.

Note: Adjustments below may in some cases cause rolling shear to control final design.

Adjustments: 0.93 for 3/8"

       1.24 for 3/8" STRUCTURAL I

       1.80 for 15/32" or 1/2" STRUCTURAL I

       1.07 for 19/32" or 5/8"

       2.37 for 19/32" or 5/8" STRUCTURAL I

       1.49 for 23/32" or 3/4"

       2.48 for 23/32" or 3/4"" STRUCTURAL I

(6) Basis: Plywood edges parallel to face grain glued to continuous framing per PDS 3.8.1

Adjustments: 1.12 for all plywood edges glued to framing per PDS 3.8.1 (1.33/1.19)

       0.84 for non-glued conditions (1.00/1.19)

Table 12.9 Effective Area of Unspliced Flange Members (PDS 5.7.3)

butt-joint spacing (t = lamination thickness)	effective laminae area
30t	90%
20t	80%
10t	60%

D. Plywood Webs

The major stress in the plywood web is through-the-panel shear stress. When calculating the maximum value of the shear stress (at the neutral axis), the standard equation, VQ/Ib, is used. In this calculation, however, the Q and I section properties are calculated including all longitudinal fibers, regardless of butt splicing. The thickness, b, is the sum of all shear thicknesses of plywood present at the section.

E. Flange to Web Connection

The shear connection between the web(s) and flange(s) causes rolling shear in the plywood. The shear stress is evaluated at the glue line with the standard equation. This stress is compared with an allowable rolling shear from Table 12.5. The table value should be reduced by 50% to account for the stress concentrations that arise at the connection (see PDS 3.3 and 3.8.2).

F. Deflections

The PDS *Supplement Two* contains a refined way to calculate the separate bending and shear components of deflection in plywood-lumber beams. An approximate method is to calculate the bending component with standard deflection equations and increase it to account for shear deflection. The magnitude of the increase is a function of the span-to-depth ratio, as indicated in Table 12.10. The values of Table 12.10 may be linearly interpolated for actual span-to-depth ratios.

Table 12.10 Bending Deflection Increase to Account for Shear (PDS Supplement Two 7.1)

span/depth	factor
10	1.5
15	1.2
20	1.0

Used with permission of APA, the Engineered Wood Association, Tacoma, WA.

G. Details

Any splices must be detailed to transmit the resultant design forces. In addition, stiffeners are added wherever a point load or support reaction is applied to the beam. These stiffeners fit vertically between the flanges to reinforce the web and distribute the point load. They are sized to not crush the flanges at the bearing points and to not induce rolling shear failure in the web as the load is transferred from flange to web.

H. Lateral Stability

Plywood-lumber beams can be long and subject to lateral buckling. The American Plywood Association suggests considering the ratio of cross-sectional stiffnesses about the vertical and horizontal axes as a measure of the beam's tendency to buckle laterally. The moments of inertia include all longitudinal fibers, regardless of splicing. The bracing recommendations, as a function of that ratio, are found in Table 12.11.

Table 12.11 Lateral Bracing Required for Plywood-Lumber Beams (PDS Supplement Two Sec. 9)

$\dfrac{\sum I_x}{\sum I_y}$	provision for lateral bracing
up to 5	none required
5 to 10	ends held in position at bottom flanges at supports
10 to 20	beams held in line at ends (both top and bottom flanges restrained from horizontal movement in planes perpendicular to beam axis)
20 to 30	one edge (either top or bottom) held in line
30 to 40	beam restrained by bridging or other bracing at intervals of not more than 8 ft
more than 40	compression flanges fully restrained and forced to deflect in a vertical plane, as with a well-fastened joist and sheathing, or stressed-skin panel system

Used with permission of APA, the Engineered Wood Association, Tacoma, WA.

Example 12.5
Plywood-Lumber Beam

A plywood-lumber beam is 24 in deep (nominally) with a 25 ft clear span. The webs are $^5/_8$ in 32/16 APA rated sheathing exterior with exterior glue. The webs are not spliced at the butt joints. The flanges are 3 douglas fir-larch, select structural 2×6s that are butt-jointed on staggered 48 in centers. The standard resurfacing has been done. The beam is subjected to normal duration and dry loading. Deflection is limited to span/360.

What is the beam's uniform load capacity as limited by (a) flange stresses, (b) web stresses, (c) web/flange connection stresses, and (d) deflection?

Solution:

The first step is to evaluate the section properties. Butt-jointed members are considered as intermittently effective in the net section properties. Once the section properties are known, the design equations are solved for the allowable uniform load.

The members are resurfaced during beam fabrication. The flange members are sanded $^1/_8$ in thinner for better gluing. After assembly, the beam's depth is resurfaced and reduced by $^1/_2$ in. This depth reduction is assumed to reduce the depth of each flange by $^1/_4$ in.

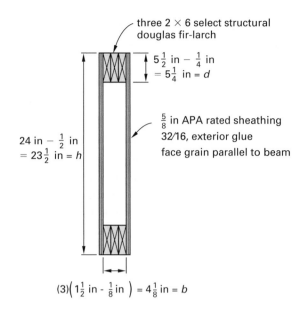

three 2 × 6 select structural
douglas fir-larch

$5\frac{1}{2}$ in $-\frac{1}{4}$ in
$= 5\frac{1}{4}$ in $= d$

$\frac{5}{8}$ in APA rated sheathing
32/16, exterior glue
face grain parallel to beam

24 in $-\frac{1}{2}$ in
$= 23\frac{1}{2}$ in $= h$

$(3)\left(1\frac{1}{2}\text{ in} - \frac{1}{8}\text{ in}\right) = 4\frac{1}{8}$ in $= b$

A. Section properties

Moments of inertia, I (net and total), are as follows.

$$\text{net } I_{\text{flanges}} = \frac{b(h^3 - (h - 2d)^3)}{12} \quad \begin{bmatrix} \text{APA PDS Supp.} \\ \text{Two App. A4.} \end{bmatrix}$$

$$= \frac{\begin{array}{c}(0.90)(2)(1.5 - 0.125) \\ \times \left((23.5)^3 - (23.5 - (2)(5.25))^3\right)\end{array}}{12}$$

$$= 2224 \text{ in}^4$$

This is the net moment of inertia for the flanges, where one of the laminae is neglected (leaving two pieces) and the remaining members are only 90% effective. The butt-joint spacing, 48 in, is 48 in/1.5 in $\approx 30t$ and Table 12.9 shows the 90%. The total moment of inertia includes all parallel fibers, regardless of butt splicing.

$$\text{total } I_{\text{flanges}} = \frac{(2224 \text{ in}^4)\left(\dfrac{3 \text{ pieces}}{2 \text{ pieces}}\right)}{0.9}$$

$$= 3707 \text{ in}^4$$

$$I_{\text{each web}} = t_{\|}\left(\frac{h^3}{12}\right)$$

$$= (0.194 \text{ in})\left(\frac{(23.5 \text{ in})^3}{12}\right)$$

$$= 210 \text{ in}^4 \quad [\text{per web}]$$

Note that from Table 12.3 [PDS Table 1], Col. 4, Unsanded Panel $^5/_8$ in,

$$t_{\|} = \frac{2.33 \text{ in}^2}{12 \text{ in}} = 0.194 \text{ in}$$

Only the plywood fibers that are parallel to the beam axis contribute to section properties. The effective thickness, $t_{\|}$, is derived from the area of parallel fibers per foot of width, 2.33 in^2/ft. Table 12.2 gives the group 4 species, and Table 12.4b gives the unsanded finish and S-1 stress rating (APA rated sheathing exterior).

The composite section properties are a combination of the plywood webs and the lumber flanges.

$$I_n = I_{\text{net}} = I_{\text{flanges}} + I_{\text{web}} \quad [\text{contributing web}]$$
$$= 2224 \text{ in}^4 + 210 \text{ in}^4 = 2434 \text{ in}^4$$
$$[\text{only one web is considered}]$$

$$I_t = I_{\text{flanges}} + I_{\text{all parallel web plys}}$$
$$= I_{\text{total}} = 3707 \text{ in}^4 + (4)(210 \text{ in}^4) = 4547 \text{ in}^4$$

The statical moment, Q_{total}, is found as follows.

$$Q_{\text{flanges}} = bd\left(\frac{h}{2} - \frac{d}{2}\right)$$
$$= \left((3)(1.5 \text{ in} - 0.125 \text{ in})\right)(5.25 \text{ in})$$
$$\times \left(\frac{23.5 \text{ in}}{2} - \frac{5.25 \text{ in}}{2}\right)$$
$$= 197.6 \text{ in}^3$$

$$Q_{\text{webs}} = t_{\|}\left(\frac{h}{2}\right)\left(\frac{h}{4}\right) \text{ (no. of webs)}$$
$$= \left(2.33 \frac{\text{in}^2}{\text{ft}}\right)\left(\frac{1 \text{ ft}}{12 \text{ in}}\right)$$
$$\times \left(\frac{23.5 \text{ in}}{2}\right)\left(\frac{23.5 \text{ in}}{4}\right)(4)$$
$$= 53.6 \text{ in}^3$$

$$Q_{\text{total}} = Q_{\text{flange}} + Q_{\text{webs}}$$
$$= 197.6 + 53.6$$
$$= 251.2 \text{ in}^3$$

Note that the only Q used is the total value. Therefore, a net value of Q is not calculated. In both I and Q calculations, the web contribution is much smaller than that of the flanges. This is one justification for not correcting for the different stiffnesses of the lumber flanges and plywood webs before combining them in their composite section.

B. Allowable loads

a) *Allowable load, limited by bending stresses in the flanges*: The allowable tensile stress for the lumber flanges is

$$F'_t = F_t(C_D C_M)$$
$$= (1000)(1.0)(1.0)$$
$$= 1000 \text{ lbf/in}^2$$

This controls over the allowable stress in the compression flange, F'_c, 1700 lbf/in^2 (see NDS Supp. Table 4A). F_t is further reduced by 20% at the butt-jointed sections (see PDS 5.7.3). The allowable moment is

$$\frac{F_t I_n}{0.5h} = \frac{(0.8)\left(1000 \dfrac{\text{lbf}}{\text{in}^2}\right)(2434 \text{ in}^4)}{(0.5)(23.5 \text{ in})}$$
$$= 165{,}719 \text{ in-lbf}$$
$$= 13{,}810 \text{ ft-lbf}$$

The allowable load is

$$\frac{M_{\text{allow}}(8)}{(\text{span})^2} = \frac{(13{,}810 \text{ ft-lbf})(8)}{(25 \text{ ft})^2}$$
$$= 176.8 \text{ lbf/ft}$$

b) *Allowable load, limited by shear stress in web*: The *Plywood Design Specification* (Sec. 3.8.1) allows a 33% increase in allowable shear through the thickness of plywood if the plywood panel is rigidly glued to continuous framing around its edges. Table 12.5 gives F_v as 130 lbf/in^2 for S-1, species group 4, dry use. For this application,

$$F_v = (1.33)(130) = 172.9 \text{ lbf/in}^2$$

The allowable horizontal shear is

$$t_s = 0.319 \text{ in} \quad [\text{from Table 12.3}]$$

$$V_h = \frac{F_v I_t t_s}{Q}$$
$$= \frac{\left(172.9 \dfrac{\text{lbf}}{\text{in}^2}\right)(4547 \text{ in}^4)\left(0.319 \dfrac{\text{in}}{\text{web}}\right)(4 \text{ webs})}{251.2 \text{ in}^3}$$
$$= 3994 \text{ lbf}$$

Neglecting the uniform load within a beam depth of the supports, the allowable uniform load for this allowable horizontal shear is

$$w_{\text{allow}} = \frac{2V}{L - 2h}$$
$$= \frac{(2)(3994 \text{ lbf})}{25 \text{ ft} - \dfrac{(2)(23.5 \text{ in})}{12 \dfrac{\text{in}}{\text{ft}}}}$$
$$= 379 \text{ lbf/ft}$$

c) *Allowable load, limited by rolling shear at the web/ flange connection*: From Table 12.5, for S-1 dry use conditions, the rolling shear at the glue line between web and flange, V_s, is found as follows.

$$F_s = 53 \text{ lbf/in}^2$$

$$V_s = \frac{2F_s d I_t}{Q_{\text{flanges}}}$$
$$= \frac{(2)\left(\dfrac{53 \dfrac{\text{lbf}}{\text{in}^2}}{2}\right)(5.25 \text{ in})(4547 \text{ in}^4)}{197.6 \text{ in}^3}$$
$$= 6403 \text{ lbf}$$

F_s is reduced 50% for the flange-to-web joints in box beams (PDS 3.8.2).

This allowable shear translates into the following allowable uniform load.

$$w_{\text{allow}}\left(\frac{V_s}{V_h}\right) = \left(379 \dfrac{\text{lbf}}{\text{ft}}\right)\left(\dfrac{6403 \text{ lbf}}{3994 \text{ lbf}}\right)$$
$$= 608 \text{ lbf/ft}$$

d) *Allowable load, limited by deflection*: A simple method of calculating deflections in plywood-lumber beams is to increase calculated bending deflections by a factor to account for the neglected, but significant, shear deflections. Table 12.10 gives this factor as a function of the span-to-depth ratio.

$$\frac{\text{span}}{\text{depth}} = \frac{25 \text{ ft}}{\dfrac{23.5 \text{ in}}{12 \dfrac{\text{in}}{\text{ft}}}}$$
$$= 12.8$$

Interpolating between ratios of 10 and 15 yields a shear deflection factor of 1.33.

The allowable deflection, span/360, is equal to $(25)(12)/360 = 0.83$ in. Use the maximum bending deflection equation for uniform load on simple spans to calculate the allowable uniform load.

$$\Delta_{\text{allow}} = \left(\frac{5w_{\text{allow}} l^4}{384 E I_t}\right)(\text{shear deflection factor})$$

$$w_{\text{allow}} = \frac{\Delta_{\text{allow}}(384) E I_t}{(\text{shear deflection factor})(5) l^4}$$
$$= \frac{(0.83 \text{ in})(384)\left(1{,}900{,}000 \dfrac{\text{lbf}}{\text{in}^2}\right)(4547 \text{ in}^4)}{(1.33)(5)(25 \text{ ft})^4\left(1728 \dfrac{\text{in}^3}{\text{ft}^3}\right)}$$
$$= 613 \text{ lbf/ft}$$

e) Maximum allowable load: The uniform load capacity of the plywood-lumber beam is limited by bending stresses to 176.8 lbf/ft and by shear in the plywood webs to 379 lbf/ft. There are several ways to increase the capacity: use thicker plywood; use a higher grade plywood—such as structural 1, with its species group 1 allowable stresses; or simply use more webs of the same plywood at the beam ends where shear stresses are maximum. The flanges could also be increased in number or size, or a higher stress species or grade could be specified.

6. Other Nonplywood Structural Panels

These panels may be broadly divided into two categories: particle boards and composite plywood panels.

A. Particle Boards

Particle boards are nonveneer panels consisting of wood particles in the form of flakes, strands, wafers, chips, shavings, and so on, that are bonded with resins under heat and pressure.

Oriented Strand Board

This is a nonveneer panel consisting of strand-like particles that are compressed and bonded with resin. Three to five layers of the strand fibers in a panel are arranged at right angles to one another in the same manner as in plywood. These panels may be rated under the APA performance specifications for structural sheathings.

Figure 12.6 Oriented Strand Board

Waferboard

This is a nonveneer structural panel that is manufactured with wood flakes (wafer-like particles) of $1\frac{1}{4}$ in or longer lengths. These flakes are randomly arranged and compressed and bonded with an adhesive, although they can be arranged in certain manners to increase the

strength and stiffness properties of a panel. These panels are rated by the Structural Board Association of Canada.

Figure 12.7 Waferboard

B. Composite Plywood Panel

The panel has a veneer face and back with an inner corestock of oriented strand material or veneer crossband. The grain direction of face and back veneers is in the long direction of the panel. These panels are rated under the APA performance specifications.

Figure 12.8 Composite Plywood Panel

Practice Problems

Practice Problem 1: Timber Formwork and Shoring

Timber formwork and shoring support a cast-in-place concrete slab during curing. The 9 in concrete slab is placed on the ³⁄₄ in, APA B-B Plyform Class II (Table 12.4b) supports that are supported by 2×4 joists spaced 16 in apart as shown. 4×6 stringers spaced 36 in apart support these joists. The shores directly under the stringers are 4×4 and are spaced at 5 ft intervals along the stringers. The lumber is no. 2 douglas fir-larch. The concrete has a specific weight of 150 lbf/ft³ and a construction live load of 50 lbf/ft². Limit all deflections to $L/360$.

Using the *National Design Specification for Wood Construction* (NDS) and the *Plywood Design Specification* by the American Plywood Association (PDS), or Chap. 12, Plywood, of this book,

1. Check the joist spacing of 16 in.
2. Check the stringer spacing of 36 in.
3. Check the shore spacing of 5 ft along each stringer.
4. Determine the allowable shore capacity.
5. Check bearing stresses at the joints connecting shore to stringer and also connecting stringer to joist.

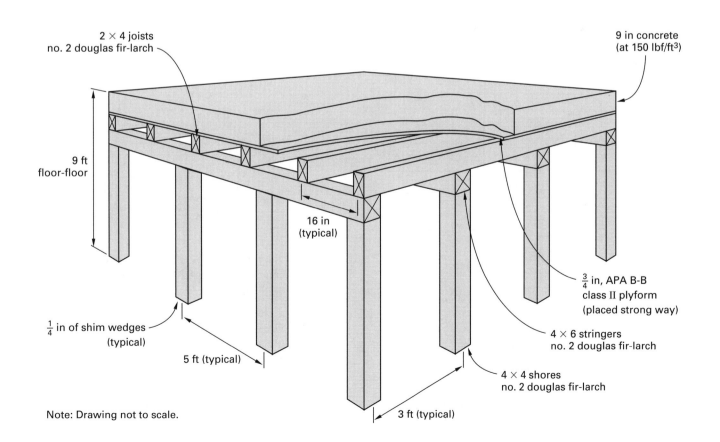

2 × 4 joists
no. 2 douglas fir-larch

9 in concrete
(at 150 lbf/ft³)

9 ft
floor-floor

16 in
(typical)

¾ in, APA B-B
class II plyform
(placed strong way)

¼ in of shim wedges
(typical)

5 ft (typical)

4 × 6 stringers
no. 2 douglas fir-larch

4 × 4 shores
no. 2 douglas fir-larch

3 ft (typical)

Note: Drawing not to scale.

Solution:

Reference

A. Joist spacing

Chap. 12

The total applied load, w, is found as follows.

$$\text{concrete slab} = \left(150 \ \frac{\text{lbf}}{\text{ft}^3}\right)\left(\frac{9 \text{ in}}{12 \text{ in}}\right)$$

$$= 112 \text{ lbf/ft}^2$$

$$\text{construction live load} = \ 50 \text{ lbf/ft}^2$$

$$\text{estimated dead loads} = \ \ 4 \text{ lbf/ft}^2$$

$$\overline{\quad\quad\quad\quad\quad\quad\quad\quad}$$

$$w \ = 166 \text{ lbf/ft}^2$$

w corresponds to 166 lbf/ft per 1 ft width of plyform board.

Check joist spacing (i.e., determine the maximum allowable plyform span length).

It is necessary to check the plyform for shear, bending, and deflection. The checks are made on 1 ft widths of the plyform.

Chap. 12,
Table 12.4b

Class II plyform is in the group 3 species, S-2 grade stress level and sanded.

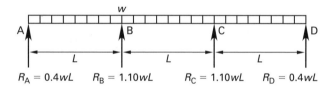

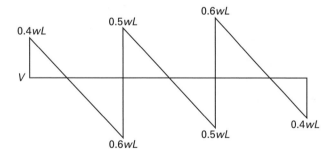

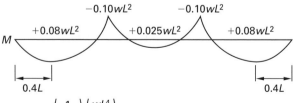

$$\Delta_{\max} = \left(\frac{1}{145}\right)\left(\frac{wL^4}{EI}\right) \text{ @ } 0.446L \text{ from A or D}$$

*Figure 13.1 Shear and Moment Diagrams
for a Three-Span Beam*

Assume that the plyform is supported in the strong direction, which is along the 8 ft side of the plyform crossing the joists. Also assume that the plyform is continuous over at least three joists (in this case, the plyform is actually continuous over more than three joists).

Chap. 12,
Table 12.5

Rolling shear is $F_s = 44 \text{ lbf/in}^2$ for wet service condition.

NDS
Table 2.3.2

For a seven-day load duration, $C_D = 1.25$.

$$F_s' = F_s C_D = \left(44 \ \frac{\text{lbf}}{\text{in}^2}\right)(1.25) = 55 \text{ lbf/in}^2$$

The calculated rolling shear, f_v, is

$$f_v = \frac{VQ}{Ib} = V\left(\frac{Q}{Ib}\right)$$

Chap. 12,
Table 12.3

The rolling shear constant for sanded panel $^3/_4$-S is

$$\frac{Ib}{Q} = 6.762 \text{ in}^2$$

The maximum shear for a three-span beam is

$$V_{\max} = V = 0.60wL$$

Setting $F_s' = f_v$,

$$55 \ \frac{\text{lbf}}{\text{in}^2} = V_{\max}\left(\frac{Q}{Ib}\right)$$

$$= 0.60wL\left(\frac{1}{6.762 \text{ in}^2}\right)$$

Rearranging,

$$L = \frac{\left(55 \ \dfrac{\text{lbf}}{\text{in}^2}\right)(6.762 \text{ in}^2)}{(0.60)\left(166 \ \dfrac{\text{lbf}}{\text{ft}}\right)\left(\dfrac{1 \text{ ft}}{12 \text{ in}}\right)} = 44.8 \text{ in}$$

Therefore, the maximum allowable joist spacing as limited by rolling shear stress is 44.8 in.

Chap. 12,
Table 12.5

For wet service condition, the allowable bending stress is

$$F_b = 820 \text{ lbf/in}^2$$

$$F_b' = F_b C_D = \left(820 \ \frac{\text{lbf}}{\text{in}^2}\right)(1.25)$$

$$= 1025 \text{ lbf/in}^2$$

The maximum calculated bending stress is

$$f_b = \frac{M}{S} = \frac{0.10wL^2}{KS}$$

Chap. 12
Table 12.3 The effective section modulus for sanded panel $^3/_4$-S is $KS = 0.412 \text{ in}^3$.

Setting $F_b' = f_b$,

$$1025 \frac{\text{lbf}}{\text{in}^2} = \frac{0.01wL^2}{0.412 \text{ in}^3}$$

Rearranging,

$$L = \sqrt{\frac{\left(1025 \frac{\text{lbf}}{\text{in}^2}\right)(0.412 \text{ in}^3)}{(0.10)\left(166 \frac{\text{lbf}}{\text{ft}}\right)\left(\frac{1 \text{ ft}}{12 \text{ in}}\right)}} = 17.5 \text{ in}$$

Therefore, the maximum allowable joist spacing as limited by bending is 17.5 in.

The calculated maximum deflection for a three-span beam is given by

$$\Delta_{\max} = \frac{wL^4}{145EI}$$

The maximum allowable deflection is

$$\Delta_{\text{allow}} = \frac{L}{360}$$

Chap. 12,
Table 12.5 For group 3 species and wet service conditions, $E = 1,100,000 \text{ lbf/in}^2$.

Chap. 12,
Table 12.3 $I = 0.197 \text{ in}^4$

Setting the maximum deflection equal to the maximum allowable deflection and rearranging,

$$\Delta_{\max} = \Delta_{\text{allow}} \text{ or } \frac{wL^4}{145EI} = \frac{L}{360}$$

$$L = \sqrt[3]{\frac{(145)\left(1,100,000 \frac{\text{lbf}}{\text{in}^2}\right)(0.197 \text{ in}^4)}{(360)\left(166 \frac{\text{lbf}}{\text{ft}}\right)\left(\frac{1 \text{ ft}}{12 \text{ in}}\right)}}$$

$$= 18.5 \text{ in}$$

Therefore, the maximum allowable joist spacing as limited by deflection is 18.5 in.

The given joist spacing of 16 in is less than the maximum allowable spacing of 17.5 in as limited by bending.

B. Check stringer spacing (i.e., maximum allowable joist span length).

For 2×4 joists of no. 2 douglas fir-larch, the NDS values for seven days' load duration and wet service are

NDS Supp.
Table 4A
Adj.
Factors

	C_F	C_M
$F_b = 900.0 \text{ lbf/in}^2$	1.5	0.85
$F_v = 95.0 \text{ lbf/in}^2$		0.97
$E = 1,600,000.0 \text{ lbf/in}^2$		0.90

NDS
Table 2.3.2 For seven days, $C_D = 1.25$.

The allowable bending stress, F_b', is

NDS
Table 2.3.1
$$F_b' = F_b C_D C_M C_L C_F C_r$$
$$= \left(900.0 \frac{\text{lbf}}{\text{in}^2}\right)(1.25)(0.85)(1.0)(1.5)(1.0)$$
$$= 1434.3 \text{ lbf/in}^2$$

Since the joist span is only 16 in, $C_L = 1.0$. Use $C_r = 1.0$ (conservative) since load-sharing among joists probably does not occur from simultaneous loading on all joists with fresh concrete.

Simplification from "point loads" to distributed loading is assumed.

The maximum calculated bending stress for a three-span beam is

$$f_b = \frac{M}{S} = \frac{0.10wL^2}{S}$$

The section modulus for 2×4 joists is

$$S = \frac{bd^2}{6} = \frac{(1.5 \text{ in})(3.5 \text{ in})^2}{6} = 3.062 \text{ in}^3$$

The joist load is

$$w = \left(166 \frac{\text{lbf}}{\text{ft}} \text{ per ft width}\right)(16 \text{ in})\left(\frac{1 \text{ ft}}{12 \text{ in}}\right)$$
$$= 221 \text{ lbf/ft}$$

Setting $F_b' = f_b$ and rearranging,

$$1434.3 \frac{\text{lbf}}{\text{in}^2} = \frac{0.10wL^2}{S}$$

$$L = \sqrt{\frac{\left(1434.3 \frac{\text{lbf}}{\text{in}^2}\right)(3.062 \text{ in}^2)}{(0.10)\left(221 \frac{\text{lbf}}{\text{ft}}\right)\left(\frac{1 \text{ ft}}{12 \text{ in}}\right)}} = 48.8 \text{ in}$$

The allowable shear stress is

NDS
Table 2.3.1

$$F'_v = F_v C_D C_M$$
$$= \left(95.0 \ \frac{\text{lbf}}{\text{in}^2}\right)(1.25)(0.97)$$
$$= 115.2 \ \text{lbf/in}^2$$

The maximum calculated shear stress for a three-span beam is

$$f_v = \frac{3V}{2A} = \frac{3(0.6wL)}{2A}$$

The cross-sectional area is

$$A = (1.5 \ \text{in})(3.5 \ \text{in}) = 5.25 \ \text{in}^2$$

Setting $F'_v = f_v$ and rearranging,

$$115.2 \ \frac{\text{lbf}}{\text{in}^2} = \frac{3(0.6wL)}{2A}$$

$$L = \frac{(2)(5.25 \ \text{in}^2)\left(115.2 \ \dfrac{\text{lbf}}{\text{in}^2}\right)}{(3)(0.6)\left(221 \ \dfrac{\text{lbf}}{\text{ft}}\right)\left(\dfrac{1 \ \text{ft}}{12 \ \text{in}}\right)}$$

$$= 36.5 \ \text{in}$$

The maximum calculated deflection for a three-span beam is

$$\Delta_{\max} = \frac{wL^4}{145E'I}$$

$$I = \frac{(1.5 \ \text{in})(3.5 \ \text{in})^3}{12} = 5.359 \ \text{in}^4$$

$$E' = EC_M = \left(1,600,000 \ \frac{\text{lbf}}{\text{in}^2}\right)(0.90)$$
$$= 1,440,000 \ \text{lbf/in}^2$$

The maximum allowable deflection is

$$\Delta_{\text{allow}} = \frac{L}{360}$$

Setting the maximum calculated deflection equal to the maximum allowable deflection and rearranging,

$$\Delta_{\max} = \Delta_{\text{allow}}$$
$$\frac{wL^4}{145E'I} = \frac{L}{360}$$

$$L = \sqrt[3]{\frac{145E'I}{360w}}$$

$$= \sqrt[3]{\frac{(145)\left(1,440,000 \ \dfrac{\text{lbf}}{\text{in}^2}\right)(5.359 \ \text{in}^4)}{(360)\left(221 \ \dfrac{\text{lbf}}{\text{ft}}\right)\left(\dfrac{1 \ \text{ft}}{12 \ \text{in}}\right)}}$$

$$= 55.3 \ \text{in}$$

Shear controls the maximum allowable joist span length. Therefore, the maximum allowable stringer spacing is 36.5 in.

C. Check shore spacing (i.e., maximum allowable stringer span length).

The NDS values for 4×6 stringers, which are no. 2 douglas fir-larch, are

NDS Supp.
Table 4A

	C_F	C_M
$F_b = 900.0 \ \text{lbf/in}^2$	1.30	1.00
$F_v = 95.0 \ \text{lbf/in}^2$		0.97
$E = 1,600,000.0 \ \text{lbf/in}^2$		0.90

For seven days, $C_D = 1.25$.

The section properties of the 4×6 lumber are

$$S = \tfrac{1}{6}bh^2 = \left(\frac{1}{6}\right)(3.5 \ \text{in})(5.5 \ \text{in})^2$$
$$= 17.65 \ \text{in}^3$$

$$I = \tfrac{1}{12}bh^3 = \left(\frac{1}{12}\right)(3.5 \ \text{in})(5.5 \ \text{in})^3$$
$$= 48.53 \ \text{in}^4$$

$$A = bh = (3.5 \ \text{in})(5.5 \ \text{in}) = 19.25 \ \text{in}^2$$

The stringer loading is

$$w = \left(166 \ \frac{\text{lbf}}{\text{ft}}\right)(36 \ \text{in})\left(\frac{1 \ \text{ft}}{12 \ \text{in}}\right) = 498 \ \text{lbf/ft}$$

NDS 3.3.3

The allowable bending stress (where $C_L = 1.0$) is

$$F'_b = F_b C_D C_M C_L C_F$$
$$= \left(900.0 \ \frac{\text{lbf}}{\text{in}^2}\right)(1.25)(1.0)(1.0)(1.3)$$
$$= 1462.5 \ \text{lbf/in}^2$$

The maximum calculated bending stress is

$$f_b = \frac{M}{S} = \frac{0.1wL^2}{S}$$

Setting $F_b' = f_b$ and rearranging,

$$1462.5 \; \frac{\text{lbf}}{\text{in}^2} = \frac{0.10wL^2}{S}$$

$$L = \sqrt{\frac{\left(1462.5 \; \dfrac{\text{lbf}}{\text{in}^2}\right)(17.65 \text{ in}^3)}{(0.10)\left(498 \; \dfrac{\text{lbf}}{\text{ft}}\right)\left(\dfrac{1 \text{ ft}}{12 \text{ in}}\right)}} = 78.9 \text{ in}$$

The maximum allowable shear stress is

$$F_v' = F_v C_D C_M$$

$$= \left(95.0 \; \frac{\text{lbf}}{\text{in}^2}\right)(1.25)(0.97)$$

$$= 115.2 \text{ lbf/in}^2$$

The maximum calculated shear stress for a three-span beam (where $V = 0.6wL$) is

$$f_v = \frac{3V}{2A}$$

Setting the maximum allowable shear stress equal to the maximum calculated shear stress and rearranging,

$$F_v' = f_v$$

$$115.2 \; \frac{\text{lbf}}{\text{in}^2} = \frac{3(0.6wL)}{2A}$$

$$L = \frac{(2)(19.25 \text{ in}^2)\left(115.2 \; \dfrac{\text{lbf}}{\text{in}^2}\right)}{(3)(0.6)\left(498 \; \dfrac{\text{lbf}}{\text{ft}}\right)\left(\dfrac{1 \text{ ft}}{12 \text{ in}}\right)} = 59.4 \text{ in}$$

Check deflections.

The maximum calculated deflection is

$$\Delta_{\text{max}} = \frac{wL^4}{145E'I}$$

NDS Supp.
Table 4A

$$E' = EC_M C_t = \left(1,600,000 \; \frac{\text{lbf}}{\text{in}^2}\right)(0.90)(1.0)$$

$$= 1,440,000 \text{ lbf/in}^2$$

The maximum allowable deflection is

$$\Delta_{\text{allow}} = \frac{L}{360}$$

Setting the maximum calculated deflection equal to the maximum allowable deflection and rearranging,

$$\Delta_{\text{max}} = \Delta_{\text{allow}}$$

$$\frac{wL^4}{145E'I} = \frac{L}{360}$$

$$L = \sqrt[3]{\frac{145E'I}{360w}}$$

$$= \sqrt[3]{\frac{(145)\left(1,440,000 \; \dfrac{\text{lbf}}{\text{in}^2}\right)(48.53 \text{ in}^4)}{(360)\left(498 \; \dfrac{\text{lbf}}{\text{ft}}\right)\left(\dfrac{1 \text{ ft}}{12 \text{ in}}\right)}}$$

$$= 87.9 \text{ in}$$

Shear stress limits the maximum allowable stringer span length to 59.4 in. For simplicity, use 60 in, that is, 5 ft, as the maximum allowable stringer span length.

D. Check shore capacity.

The shore load is

$$P = (\text{applied loads})(\text{area}) = wA$$

$$= \left(166 \; \frac{\text{lbf}}{\text{ft}^2}\right)((3 \text{ ft})(5 \text{ ft}))$$

$$= 2490 \text{ lbf}$$

The shore is a solid column with a total length calculated as follows.

$l = 9 \text{ ft} - 0 \text{ in}$	floor-floor	
less	9 in	slab
	$^3/_4$ in	plyform
	$3^1/_2$ in	2×4 joist
	$5^1/_2$ in	4×6 stringer
	$^1/_4$ in	shims

$$\text{total } l = 89.0 \text{ in}$$
$$l_u = l = 89.0 \text{ in}$$

For 4×4 shores of no. 2 douglas fir-larch, the NDS values for seven days' load duration and wet service are

	C_F	C_M
$F_c = 1350.0 \text{ lbf/in}^2$	1.15	0.8
$E = 1,600,000.0 \text{ lbf/in}^2$		0.90

The allowable compression design value is

$$F_c' = F_c C_D C_M C_t C_F C_P = F_c^* C_P$$

$$F_c^* = \left(1350 \; \frac{\text{lbf}}{\text{in}^2}\right)(1.25)(0.8)(1.0)(1.15)$$

$$= 1552.5 \text{ lbf/in}^2$$

$$E' = EC_M C_t = \left(1,600,000 \; \frac{\text{lbf}}{\text{in}^2}\right)(0.9)(1.0)$$

$$= 1,440,000 \text{ lbf/in}^2$$

NDS 3.7   Determine the column stability factor, C_P.

NDS App. G $K_e = 1.0$

NDS 3.7.1

$$\left(\frac{l_e}{d}\right)_x = \left(\frac{l_e}{d}\right)_y = \frac{K_e l_u}{d} = \frac{(1.0)(89.0 \text{ in})}{3.5 \text{ in}}$$

$$= 25.4$$

$K_{cE} = 0.3$ for visually graded lumber, and $c = 0.8$ for sawn lumber.

$$F_{cE} = \frac{K_{cE} E'}{\left(\frac{l_e}{d}\right)^2} = \frac{(0.3)\left(1{,}440{,}000 \dfrac{\text{lbf}}{\text{in}^2}\right)}{(25.4)^2}$$

$$= 669.6 \text{ lbf/in}^2$$

$$\frac{F_{cE}}{F_c^*} = \frac{669.6 \dfrac{\text{lbf}}{\text{in}^2}}{1552.5 \dfrac{\text{lbf}}{\text{in}^2}} = 0.431$$

$$\frac{1 + \dfrac{F_{cE}}{F_c^*}}{2c} = \frac{1 + 0.431}{(2)(0.8)} = 0.894$$

NDS Eq. 3.7-1

$$C_P = \frac{1 + \dfrac{F_{cE}}{F_c^*}}{2c} - \sqrt{\left(\frac{1 + \dfrac{F_{cE}}{F_c^*}}{2c}\right)^2 - \frac{\dfrac{F_{cE}}{F_c^*}}{c}}$$

$$= 0.894 - \sqrt{(0.894)^2 - \frac{0.431}{0.8}} = 0.384$$

The allowable compressive stress is

$$F_c' = F_c^* C_P = \left(1552.5 \frac{\text{lbf}}{\text{in}^2}\right)(0.384)$$

$$= 596.2 \text{ lbf/in}^2$$

The calculated compressive stress is

$$f_c = \frac{2490 \text{ lbf}}{(3.5 \text{ in})(3.5 \text{ in})} = 203.3 \text{ lbf/in}^2$$

$$203.3 \text{ lbf/in}^2 < 592.0 \text{ lbf/in}^2$$

$$f_c < F_c' \quad \text{[OK]}$$

The shore capacity is

$$P_{\text{allow}} = F_c'(4 \times 4 \text{ shore area})$$

$$= \left(596.2 \frac{\text{lbf}}{\text{in}^2}\right)((3.5 \text{ in})(3.5 \text{ in}))$$

$$= 7303.5 \text{ lbf} > \text{shore load of } 2490 \text{ lbf} \quad \text{[OK]}$$

NDS 2.3.10 and 3.10   **E. Check bearing stresses.**

The NDS value for compression perpendicular to grain for no. 2 douglas fir-larch is

NDS Supp. 4A

$$F_{c\perp} = 625 \text{ lbf/in}^2$$
$$C_M = 0.67$$
$$C_t = 1.0 \text{ for normal temperature}$$

The allowable compressive (bearing) stress design value is

NDS Table 2.3.1; NDS 2.3.10 and 3.10.2

$$F_{c\perp}' = F_{c\perp} C_M C_t C_b$$

C_b can be used for $l_b < 6$ in and when bearing is at least 3 in away from member end.

$$C_b = \frac{l_b + \frac{3}{8} \text{ in}}{l_b}$$

For 4×4 shores with 4×6 stringers, see Fig. 13.2.

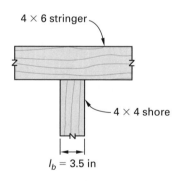

Figure 13.2 Bearing Stresses on Column

$$C_b = \frac{l_b + \frac{3}{8} \text{ in}}{l_b} = \frac{3.5 \text{ in} + \frac{3}{8} \text{ in}}{3.5 \text{ in}} = 1.107$$

$$F_{c\perp}' = \left(625 \frac{\text{lbf}}{\text{in}^2}\right)(0.67)(1.0)(1.107)$$

$$= 463.6 \text{ lbf/in}^2$$

Checking bearing stress gives

$$f_{c\perp} = \frac{\text{shore load}}{\text{bearing area}} = \frac{2490 \text{ lbf}}{(3.5 \text{ in})(3.5 \text{ in})}$$

$$= 203.3 \text{ lbf/in}^2$$

$$203.3 \text{ lbf/in}^2 < 463.6 \text{ lbf/in}^2$$

$$f_{c\perp} < F'_{c\perp} \quad [\text{OK}]$$

For 4×6 stringer with 2×4 joist, see Fig. 13.3.

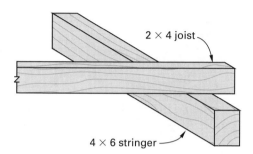

Figure 13.3 Bearing Stresses on Stringer

Since both members are compressed perpendicular to grain, the following allowable value is applicable, with a factor of $1/2$.

$$F'_{c\perp} = \left(\tfrac{1}{2}\right)\left(F_{c\perp} C_M C_t C_b\right)$$

$$C_b = \frac{l_b + \tfrac{3}{8} \text{ in}}{l_b} = \frac{3.5 \text{ in} + \tfrac{3}{8} \text{ in}}{3.5 \text{ in}} = 1.107$$

$$F'_{c\perp} = \left(\frac{1}{2}\right)\left(625 \; \frac{\text{lbf}}{\text{in}^2}\right)(0.67)(1.0)(1.107)$$

$$= 231.8 \text{ lbf/in}^2$$

The bearing load is

$$P = \left(166 \; \frac{\text{lbf}}{\text{ft}^2}\right)(16 \text{ in})(36 \text{ in})\left(\frac{1 \text{ ft}^2}{144 \text{ in}^2}\right)$$

$$= 664 \text{ lbf}$$

The actual bearing stress is

$$f_{c\perp} = \frac{P}{\text{bearing area}} = \frac{664 \text{ lbf}}{(1.5 \text{ in})(3.5 \text{ in})}$$

$$= 126.5 \text{ lbf/in}^2$$

$$126.5 \text{ lbf/in}^2 < 231.8 \text{ lbf/in}^2$$

$$f_{c\perp} < F_{c\perp} \quad [\text{OK}]$$

Therefore, the shores have sufficient load capacity.

Practice Problem 2: Bridge Stringers and Deck

A section through the deck of a 30 ft long single-span timber bridge is shown.

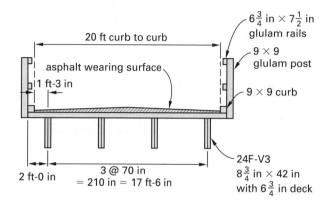

The following data apply.

- Width: 20 ft curb-to-curb
- Span: 30 ft
- Asphalt wearing surface average thickness: 3 in
- Deck: glulam douglas fir 22F-E1, $6^3/4$ in thick with bonded edge joint
- Stringers: glulam douglas fir 24F-V4, $8^3/4$ in $\times$ 42 in
- Applicable specification: AASHTO 1996
- Live load: HS20-44
- Timber unit weight: 50 lbf/ft^3 (AASHTO 3.3.6)
- Allowable live load deflection: $L/500$

The maximum dead load plus live load moments are 221.84 ft-k for each interior stringer and 12.25 ft-k for a 21.75 in deck width.

The maximum dead plus live load shears in kips are 27.763 k for each interior stringer and 7.422 k for a 21.75 in deck width.

Check whether the interior stringers and deck are sufficiently structurally designed. Assume that the deck-stringer connections provide continuous lateral support to the stringers. Checking deck deflection is not required for this problem.

Solution:

Reference

A. Check interior stringers.
($8^3/4$ in $\times$ 42 in, glulam douglas fir 24F-V4)

NDS Supp.   The section properties of an $8^3/4$ in $\times$ 42 in
Table 1C   glulam beam are

$$A = 367.5 \text{ in}^2$$
$$S_x = 2572.5 \text{ in}^3$$
$$I_x = 54{,}022.5 \text{ in}^4$$

NDS Supp.
Table 5A

Design values for glulam combination 24F-V4 douglas fir are

$$
\begin{array}{ll}
& C_M \\
\hline
F_{bx} = 2400 \ \text{lbf/in}^2 & 0.80 \\
F_{vx} = 190 \ \text{lbf/in}^2 & 0.875 \\
E_x = 1{,}800{,}000 \ \text{lbf/in}^2 & 0.833
\end{array}
$$

NDS Supp.
Table 5A
Adj. Factors
or NDS 5.3.2

The volume factor is

$$
C_V = K_L \left(\frac{21}{L}\right)^{1/x} \left(\frac{12}{d}\right)^{1/x} \left(\frac{5.125}{b}\right)^{1/x}
$$
$$
\leq 1.0
$$

$K_L = 1.09$ for concentrated load at midspan, and $x = 10$ for other than southern pine.

Thus,

$$
C_V = (1.09) \left(\frac{21}{30 \ \text{ft}}\right)^{1/10} \left(\frac{12}{42 \ \text{in}}\right)^{1/10}
$$
$$
\times \left(\frac{5.125}{8.75 \ \text{in}}\right)^{1/10}
$$
$$
= 0.8796
$$

NDS 3.3.3.3

The beam stability factor, C_L, is 1.0 since the compression edge of the stringer is continuously laterally supported by the deck, assuming that the deck is fastened appropriately. Therefore, $C_V = 0.8796$ controls.

NDS
Table 2.3.2

$C_D = 1.0$ for 10-year load duration.

NDS
Table 2.3.1

The allowable design values are

$$
\begin{aligned}
F_b' &= F_b C_D C_M C_V \\
&= \left(2400 \ \frac{\text{lbf}}{\text{in}^2}\right)(1.0)(0.8)(0.8796) \\
&= 1688.8 \ \text{lbf/in}^2
\end{aligned}
$$

$$
\begin{aligned}
F_v' &= F_v C_D C_M \\
&= \left(190 \ \frac{\text{lbf}}{\text{in}^2}\right)(1.0)(0.875) \\
&= 166.3 \ \text{lbf/in}^2
\end{aligned}
$$

$$
\begin{aligned}
E' &= E_x C_M \\
&= \left(1{,}800{,}000 \ \frac{\text{lbf}}{\text{in}^2}\right)(0.833) \\
&= 1{,}499{,}400 \ \text{lbf/in}^2
\end{aligned}
$$

Check the bending stress.

$$
M_{\max} = 221.84 \ \text{ft-k} \quad \text{[given]}
$$

The calculated bending stress is

$$
\begin{aligned}
f_b &= \frac{M_{\max}}{S_x} \\
&= \left(\frac{221.84 \ \text{ft-k}}{2572.5 \ \text{in}^3}\right)\left(12 \ \frac{\text{in}}{\text{ft}}\right)\left(1000 \ \frac{\text{lbf}}{\text{k}}\right) \\
&= 1034.8 \ \text{lbf/in}^2 \\
& \quad 1034.8 \ \text{lbf/in}^2 < 1688.8 \ \text{lbf/in}^2 \\
& \qquad\qquad f_b < F_b' \quad \text{[OK]}
\end{aligned}
$$

Check the shear stress.

$$
V_{\max} = 27.763 \ \text{k} \quad \text{[given]}
$$

The calculated shear stress is

$$
\begin{aligned}
f_v &= \frac{3 V_{\max}}{2A} \\
&= \left(\frac{(3)(27.763 \ \text{k})}{(2)(367.5 \ \text{in}^2)}\right)\left(1000 \ \frac{\text{lbf}}{\text{k}}\right) \\
&= 113.3 \ \text{lbf/in}^2 \\
& \quad 113.3 \ \text{lbf/in}^2 < 166.3 \ \text{lbf/in}^2 \\
& \qquad\qquad f_v < F_v' \quad \text{[OK]}
\end{aligned}
$$

Check the live load deflection.

$$
\Delta = \frac{PL^3}{48 E' I}
$$

AASHTO
Table 3.23.1

The wheel load multiplied by the stringer distribution factor is

$$
P = (16{,}000 \ \text{lbf})(1.167) = 18{,}672 \ \text{lbf}
$$

$$
\Delta = \frac{(18{,}672 \ \text{lbf})(360 \ \text{in})^3}{(48)\left(1{,}499{,}400 \ \frac{\text{lbf}}{\text{in}^2}\right)(54{,}022.5 \ \text{in}^4)}
$$
$$
= 0.224 \ \text{in}
$$

$$
\Delta_{\text{allow}} = \frac{L}{500} = \frac{360 \ \text{in}}{500} = 0.72 \ \text{in}
$$
$$
0.72 \ \text{in} > 0.224 \ \text{in}
$$
$$
\Delta_{\text{allow}} > \Delta \quad \text{[OK]}
$$

Use $8^3/_4$ in $\times$ 42 in glulam, 24F-V4 douglas fir stringer.

B. Check the deck thickness ($6^3/_4$ in glulam douglas fir 22F-E1).

AASHTO
3.25.1

The $6^3/_4$ in thick section properties are based on a 21.75 in deck width and a 68 in deck span.

$$
A = (21.75 \ \text{in})(6.75 \ \text{in}) = 146.81 \ \text{in}^2
$$
$$
S_x = \frac{(21.75 \ \text{in})(6.75 \ \text{in})^2}{6} = 165.16 \ \text{in}^3
$$

NDS Supp.
Tables 5A
and 5A Adj.
Factors

Design values for glulam combination 22F-E1 douglas fir are

	C_M
$F_{by} = 1550 \ \text{lbf/in}^2$	0.80
$F_{vy} = 165 \ \text{lbf/in}^2$	0.875
$E_y = 1,600,000 \ \text{lbf/in}^2$	0.833
$C_{fu} = $ flat use factor	1.07 for $6^3/4$ in thickness
$C_D = $	1.0 for 10-year load duration

NDS 3.3.3
and 5.3.2

Volume factor, C_V, and beam stability factor, C_L, are both equal to 1.0 because of the deck thickness and geometry.

The allowable design values are

$$F_b' = F_{by} C_D C_M C_{fu} \left(\begin{array}{c} \text{the smaller of} \\ C_L \text{ and } C_V \end{array} \right)$$

$$= \left(1550 \ \frac{\text{lbf}}{\text{in}^2} \right) (1.0)(0.80)(1.07)(1.0)$$

$$= 1326.8 \ \text{lbf/in}^2$$

$$F_v' = F_{vy} C_D C_M = \left(165 \ \frac{\text{lbf}}{\text{in}^2} \right) (1.0)(0.875)$$

$$= 144.4 \ \text{lbf/in}^2$$

Check bending and shear stresses.

$$f_b = \frac{M}{S_x} = \left(\frac{12.25 \ \text{ft-k}}{165.16 \ \text{in}^3} \right) \left(12 \ \frac{\text{in}}{\text{ft}} \right) \left(1000 \ \frac{\text{lbf}}{\text{k}} \right)$$

$$= 890.0 \ \text{lbf/in}^2$$

$$890.0 \ \text{lbf/in}^2 < 1326.8 \ \text{lbf/in}^2$$

$$f_b < F_b' \quad [\text{OK}]$$

$$f_v = \frac{3V}{2A} = \frac{(3)(7.422 \ \text{k}) \left(1000 \ \frac{\text{lbf}}{\text{k}} \right)}{(2)(146.81 \ \text{in}^2)}$$

$$= 75.8 \ \text{lbf/in}^2$$

$$75.8 \ \text{lbf/in}^2 < 144.4 \ \text{lbf/in}^2$$

$$f_v < F_v' \quad [\text{OK}]$$

Use $6^3/4$ in thick glulam 22F-E1 deck.

(Note: A deflection check for this deck is not necessary for this problem.)

Practice Problem 3: Commercial Building

Snow loading and wind pressures on a large one-story commercial building have been determined in accordance with ASCE 7-95 provisions. The design loads are dead loads plus snow loads for the roof structure and dead loads plus wind loads for the columns. The design snow load is 20 lbf/ft^2 and the wind pressure on the windward wall is 10 lbf/ft^2. The roof structure consists of a $1\frac{1}{2}$ in thick timber plank deck supported by purlins spaced at 8 ft on centers. The purlins rest on frame rafters spaced at 30 ft on centers as shown. The maximum allowable total deflection is $L/180$. Assume timber unit weights of 36 lbf/ft^3 for the deck and 40 lbf/ft^3 for the other timber components. Use douglas fir-larch.

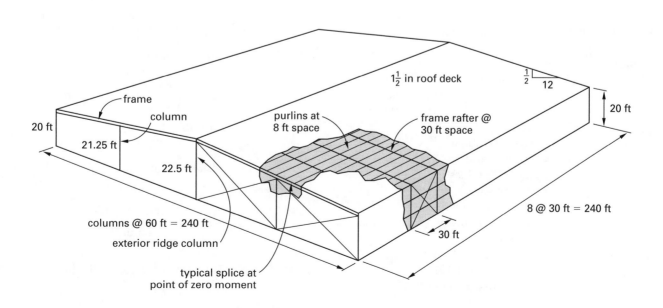

1. Design the deck.
2. Design the purlins.
3. Design the frame rafters (assume two-span cantilever beam system).
4. Design the connections between the purlins and frame rafters.
5. Design the end column.

Solution:

Reference

A. Design roof deck.

The deck span is 8 ft between roof purlins and is continuous over two purlins. The deck consists of 2 × 6 tongue-and-groove douglas fir-larch planks, commercial dex grade.

The total roof load, w, is found as follows.
Find dead load.

roofing, felt, and
vapor barrier $= 5.0$ lbf/ft^2

insulation $= 4.0$ lbf/ft^2

2 in nominal tongue-
and-groove plank $= \left(36 \dfrac{\text{lbf}}{\text{ft}^3}\right)(1.5 \text{ in})\left(\dfrac{1 \text{ ft}}{12 \text{ in}}\right)$

$= 4.5$ lbf/ft^2

$\overline{\hspace{3cm}}$

DL $= 13.5$ lbf/ft^2

Find snow load.

SL $= 20.0$ lbf/ft^2

$\overline{\hspace{3cm}}$

The total load, w, is

DL + SL $= 13.5 \dfrac{\text{lbf}}{\text{ft}^2} + 20.0 \dfrac{\text{lbf}}{\text{ft}^2} = 33.5$ lbf/ft^2

NDS Supp.   Design values are as follows.
Table 4E, 4E
Adj. Factors

		C_F
F_b	= 1450 lbf/in^2	$\overline{\hspace{1cm}}$ 1.10 for 2 in nominal thickness
$F_{c\perp}$	= 625 lbf/in^2	
E	= 1,700,000 lbf/in^2	
F_v	= not available in the NDS tables	
C_{fu}	= flat use factor already considered in the NDS design values	

NDS Table 2.3.2 C_D = 1.15 for snow loads
C_M = 1.0 for 15% or less moisture content

NDS
Table 2.3.1

Allowable design values are

$F_b' = F_b C_D C_M C_F$

$= \left(1450 \dfrac{\text{lbf}}{\text{in}^2}\right)(1.15)(1.0)(1.10)$

$= 1834.2$ lbf/in^2

$F_{c\perp}' = F_{c\perp} C_M = \left(625 \dfrac{\text{lbf}}{\text{in}^2}\right)(1.0)$

$= 625$ lbf/in^2

$E' = E C_M = \left(1,700,000 \dfrac{\text{lbf}}{\text{in}^2}\right)(1.0)$

$= 1,700,000$ lbf/in^2

2 in nominal deck section properties per foot width are

$A = (1.5 \text{ in})(12 \text{ in}) = 18.0 \text{ in}^2$

$S = \dfrac{bh^2}{6} = \dfrac{(12 \text{ in})(1.5 \text{ in})^2}{6} = 4.5 \text{ in}^3$

$I = \dfrac{bh^3}{12} = \dfrac{(12 \text{ in})(1.5 \text{ in})^3}{12} = 3.375 \text{ in}^4$

The load per foot on the 2 in nominal deck is

$\left(33.5 \dfrac{\text{lbf}}{\text{ft}^2}\right)(1 \text{ ft}) = 33.5$ lbf/ft

The maximum moment and shear for a two-span deck are

$M = -\dfrac{1}{8}wL^2$

$= -\left(\dfrac{1}{8}\right)\left(33.5 \dfrac{\text{lbf}}{\text{ft}}\right)(8 \text{ ft})^2\left(12 \dfrac{\text{in}}{\text{ft}}\right)$

$= -3216.0$ in-lbf

$V = \dfrac{5}{8}wL = \left(\dfrac{5}{8}\right)\left(33.5 \dfrac{\text{lbf}}{\text{ft}}\right)(8 \text{ ft})$

$= 167.5$ lbf

The actual maximum bending and shear stresses are

$f_b = \dfrac{|M|}{S} = \dfrac{|-3216.0 \text{ in-lbf}|}{4.5 \text{ in}^3}$

$= 714.7$ lbf/in^2

$714.7 \text{ lbf/in}^2 < 1834.2 \text{ lbf/in}^2$

$f_b < F_b'$ [OK]

$$f_v = \frac{3V}{2A} = \frac{(3)(167.5 \text{ lbf})}{(2)(18.0 \text{ in}^2)}$$

$$= 13.96 \text{ lbf/in}^2 \quad \begin{bmatrix} \text{very small by} \\ \text{inspection; OK} \end{bmatrix}$$

Check the maximum deflection for the 2 in nominal deck per foot width.

$$\Delta = \frac{wL^4}{185E'I}$$

$$= \frac{\left(33.5 \dfrac{\text{lbf}}{\text{ft}}\right)\left(\dfrac{1 \text{ ft}}{12 \text{ in}}\right)\left((8 \text{ ft})\left(12 \dfrac{\text{in}}{\text{ft}}\right)\right)^4}{(185)\left(1,700,000 \dfrac{\text{lbf}}{\text{in}^2}\right)(3.375 \text{ in}^4)}$$

$$= 0.223 \text{ in}$$

$$\Delta_{\text{allow}} = \frac{L}{180} = \frac{(8 \text{ ft})\left(12 \dfrac{\text{in}}{\text{ft}}\right)}{180} = 0.53 \text{ in}$$

$$0.53 \text{ in} > 0.223 \text{ in}$$

$$\Delta_{\text{allow}} > \Delta \quad [\text{OK}]$$

Use 2 in nominal deck, douglas fir-larch commercial dex, continuous two-span, 8 ft each span.

B. Design purlins.

Assume glulam $3^1/_8$ in $\times$ 18 in, 20F-V3 douglas fir at 8 ft spacing. Assume timber density of 40 lbf/ft^3. Alternatively, calculate timber densities from specific gravities in NDS Tables in NDS Parts 8, 9, 11, and 12.

The total purlin load, w, is found as follows.

$$\begin{aligned} \text{roof} \\ \text{loads} \end{aligned} = \left(33.5 \frac{\text{lbf}}{\text{ft}^2}\right)(8 \text{ ft})$$

$$= 268.0 \text{ lbf/ft}$$

$$\begin{aligned} \text{purlin} \\ \text{dead} \\ \text{load} \end{aligned} = \frac{\left(40 \dfrac{\text{lbf}}{\text{ft}^3}\right)(3.125 \text{ in})(18 \text{ in})(1 \text{ ft}^2)}{144 \text{ in}^2}$$

$$= 15.6 \text{ lbf/ft}$$

$$w = 268.0 \frac{\text{lbf}}{\text{ft}} + 15.6 \frac{\text{lbf}}{\text{ft}}$$

$$= 283.6 \text{ lbf/ft}$$

NDS Supp. Table 1C — Section properties of $3^1/_8$ in $\times$ 18 in glulam are as follows.

span $=$ 30 ft simply supported
$A = 56.25 \text{ in}^2$
$S_x = 168.8 \text{ in}^3$
$I_x = 1518.8 \text{ in}^4$

NDS Supp. Table 5A — Design values for glulam combination 20F-V3 douglas fir are

$F_{bx} = 2000 \text{ lbf/in}^2$
$F_{vx} = 190 \text{ lbf/in}^2$
$E_x = 1,600,000 \text{ lbf/in}^2$
$C_M = 1.0$ for assumed 15% or less
  moisture content
$C_D = 1.15$ for snow load

NDS 3.3.3.3 — The beam stability factor, C_L, is 1.0 since the joists are continuously supported by the deck above it.

NDS Supp. Table 5A-Adj. Factors or NDS 5.3.2 — The volume factor is

$$C_V = K_L \left(\frac{21}{L}\right)^{1/x} \left(\frac{12}{d}\right)^{1/x} \left(\frac{5.125}{b}\right)^{1/x}$$
$$\leq 1.0$$

$K_L = 1.0$ for uniformly distributed load, and $x = 10.0$ for other than southern pine

$$C_V = (1.0)\left(\frac{21}{30 \text{ ft}}\right)^{1/10} \left(\frac{12}{18 \text{ in}}\right)^{1/10}$$

$$= \quad \times \left(\frac{5.125}{3.125 \text{ in}}\right)^{1/10}$$

$$= 0.974$$

NDS Table 2.3.1 — The allowable design values are

$$F_b' = (F_{bx}C_DC_M)\begin{pmatrix} \text{the smaller of} \\ C_V \text{ and } C_L \end{pmatrix}$$

$$= \left(2000 \frac{\text{lbf}}{\text{in}^2}\right)(1.15)(1.0)(0.974)$$

$$= 2240.2 \text{ lbf/in}^2$$

$$F_v' = F_{vx}C_DC_M = \left(190 \frac{\text{lbf}}{\text{in}^2}\right)(1.15)(1.0)$$

$$= 218.5 \text{ lbf/in}^2$$

$$E' = EC_M = \left(1,600,000 \frac{\text{lbf}}{\text{in}^2}\right)(1.0)$$

$$= 1,600,000 \text{ lbf/in}^2$$

Check the actual maximum stresses.

$$M = \frac{wL^2}{8} = \left(\frac{\left(283.6 \; \frac{\text{lbf}}{\text{ft}}\right)(30 \text{ ft})^2}{8} \right) \left(12 \; \frac{\text{in}}{\text{ft}} \right)$$

$$= 382{,}860 \text{ in-lbf}$$

$$f_b = \frac{M}{S_x} = \frac{382{,}860 \text{ in-lbf}}{168.8 \text{ in}^3}$$
$$= 2268.1 \text{ lbf/in}^2$$
$$2268.1 \text{ lbf/in}^2 \approx 2240.2 \text{ lbf/in}^2$$
$$f_b \approx F_b' \quad \text{[consider OK]}$$

(Note: When actual values slightly exceed allowable design values, engineering judgment must be used to determine whether or not this is acceptable. However, this may not be the case when taking the engineering registration examination(s) and strict compliance may be required.)

NDS 3.4.3

$$V = w \left(\frac{\text{span}}{2} - \text{beam depth} \right)$$

$$= \left(283.6 \; \frac{\text{lbf}}{\text{in}^2} \right) \left((30 \text{ ft}) \left(\frac{1}{2} \right) - (18 \text{ in}) \left(\frac{1 \text{ ft}}{12 \text{ in}} \right) \right)$$

$$= 3828.6 \text{ lbf}$$

$$f_v = \frac{3V}{2A} = \frac{(3)(3828.6 \text{ lbf})}{(2)(56.25 \text{ in}^2)}$$
$$= 102.1 \text{ lbf/in}^2$$
$$102.1 \text{ lbf/in}^2 < 218.5 \text{ lbf/in}^2$$
$$f_v < F_v' \quad \text{[OK]}$$

Check the maximum deflection.

$$\Delta = \frac{5wL^4}{384E'I}$$

$$= \frac{(5)\left(\left(283.6 \; \frac{\text{lbf}}{\text{ft}} \right) \left(\frac{1 \text{ ft}}{12 \text{ in}} \right) \right) \left((30 \text{ ft}) \left(12 \; \frac{\text{in}}{\text{ft}} \right) \right)^4}{(384) \left(1{,}600{,}000 \; \frac{\text{lbf}}{\text{in}^2} \right) (1518.8 \text{ in}^4)}$$

$$= 2.13 \text{ in}$$

$$\Delta_{\text{allow}} = \frac{L}{180} = \frac{(30 \text{ ft}) \left(12 \; \frac{\text{in}}{\text{ft}} \right)}{180} = 2.0 \text{ in}$$
$$2.0 \text{ in} \approx 2.13 \text{ in}$$
$$\Delta_{\text{allow}} \approx \Delta \quad \text{[considered OK]}$$

(Note: When actual values slightly exceed allowable design values, engineering judgment must be used to determine whether or not this is acceptable. However, this may

not be the case when taking the engineering registration examination(s) and strict compliance may be required.)

Use glulam $3^{1}/_{8}$ in × 18 in, 20F-V3 douglas fir for the 30 ft purlins.

C. Design the frame rafter.

Assume glulam $8^{3}/_{4}$ in × 39 in 24F-V8 douglas fir and a two-span cantilever beam system. Assume a timber density of 40 lbf/ft³.

An optimum design for bending strength is determined by choosing the two-span beam system as shown in Fig. 13.4. The splice is placed at a location where there is no bending moment and is modeled as a hinge. See the Appendix: Beam Formulas or the *AITC Timber Construction Manual* for appropriate equations (Reference 3 in the Appendix to this book). Or the equations can be derived as done in this problem solution.

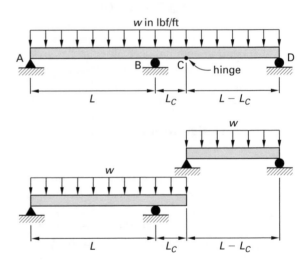

Figure 13.4 Two-Span Cantilever Beam System

The position of an inflection point, L_c, is determined by making both the negative moment at B and positive moment in span CD equal in absolute value. The negative moment at B is

$$M_{\text{B,neg}} = \frac{wL_c^2}{2} + \frac{w(L - L_c)L_c}{2}$$

The maximum positive moment in span CD is

$$M_{\text{CD,pos}} = \frac{w(L - L_c)^2}{8}$$

By letting $|M_B| = |M_{CD}|$ and rearranging,

$$\frac{wL_c^2}{2} + \frac{w(L - L_c)L_c}{2} = \frac{w(L - L_c)^2}{8}$$

Therefore, the inflection point, L_c, is located at $L_c = 0.172L$.

Also, the moment, M, is found to be $M = 0.086wL^2$.

The shears, V, are as follows.

At the ends (point A or point D),

$$V = \frac{w(L - L_c)}{2} = \frac{w(L - 0.172L)}{2}$$
$$= 0.414wL$$

At center (at B),

$$V = -0.414wL - wL_c$$
$$= -0.414wL - w(0.172L)$$
$$= -0.586wL$$

NDS Supp.
Table 1C

Section properties of $8^3/_4$ in $\times$ 39 in glulam are

span $= 60$ ft
$A = 341.2$ in^2
$S_x = 2218$ in^3
$I_x = 43{,}253$ in^4

NDS Supp.
Table 5A

Design values for glulam 24F-V8 douglas fir spaced at 30 ft with 60 ft spans are as follows.

The tension zone stressed in tension is
$F_{bx,t/t} = 2400$ lbf/in^2

The compression zone stressed in tension is
$F_{bx,c/t} = 2400$ lbf/in^2
$F_{vx} = 190$ lbf/in^2
$E_x = 1{,}800{,}000$ lbf/in^2
$E_y = 1{,}600{,}000$ lbf/in^2
$C_M = 1.0$ for dry condition
$C_D = 1.15$ for snow

The rafter load is

$$\text{roof loads (snow plus deck)} = \left(33.5 \ \frac{\text{lbf}}{\text{ft}^2}\right)(30 \text{ ft})$$
$$= 1005 \text{ lbf/ft}$$

$$\text{purlins} = \left(40 \ \frac{\text{lbf}}{\text{ft}^3}\right)\left(\frac{25}{8} \text{ in}\right)(18 \text{ in})\left(\frac{1 \text{ ft}^2}{144 \text{ in}^2}\right)$$
$$\times (30 \text{ ft})\left(\frac{1}{8 \text{ ft}}\right)$$
$$= 58.6 \text{ lbf/ft}$$

$$\text{rafter} = \left(40 \ \frac{\text{lbf}}{\text{ft}^3}\right)\left(\frac{35}{4} \text{ in}\right)(39 \text{ in})\left(\frac{1 \text{ ft}^2}{144 \text{ in}^2}\right)$$
$$= 94.80 \text{ lbf/ft}$$
$$w = 1005 \ \frac{\text{lbf}}{\text{ft}} + 58.6 \ \frac{\text{lbf}}{\text{ft}} + 94.8 \ \frac{\text{lbf}}{\text{ft}}$$
$$= 1158.4 \text{ lbf/ft} \quad (1.158 \text{ k/ft})$$

The rafter glulam 24F-V8 douglas fir is loaded with snow loads on both spans as shown in Fig. 13.5. The inflection point is located at distance L_c from B.

$$L_c = 0.172L = (0.172)(60 \text{ ft}) = 10.32 \text{ ft}$$

It is noted that this region is where the compression zone of the rafter glulam is stressed in tension. The moments and shears are shown in Fig. 13.5.

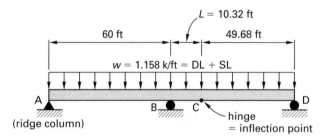

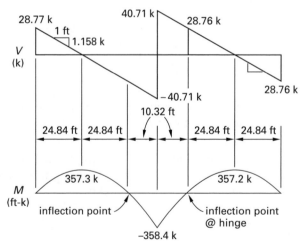

Figure 13.5 Behavior of Rafter Under Load
(Forces and moments deviate slightly as calculated from Appendix: Beam Formulas because of round-off error.)

For member AC (see Fig. 13.5):

NDS Supp.
Table 5A

Bending: glulam 24F-V8 douglas fir combination is "balanced" to provide equal positive and negative moment capacity.

$F_{bx} = 2400 \text{ lbf/in}^2$

(F_{bx} for tension zone stressed in tension = F_{bx} for compression zone stressed in tension.)

$M_{\max} = -M_{\max} = 358.4 \text{ ft-k}$
$$\begin{bmatrix} \text{see Fig. 13.5 and} \\ \text{Appendix: Beam Formulas} \end{bmatrix}$$

The volume effect factor is determined as follows.

NDS 5.3.2 $C_V = K_L \left(\dfrac{21}{L}\right)^{1/x} \left(\dfrac{12}{d}\right)^{1/x} \left(\dfrac{5.125}{b}\right)^{1/x}$
$$\leq 1.0$$

For uniformly distributed load, $K_L = 1.0$.

For other than southern pine, $x = 10.0$.

The largest distance between points of zero moments is

$L = 60 \text{ ft} - 10.32 \text{ ft} = 49.68 \text{ ft}$

NDS 5.3.2 $C_V = K_L \left(\dfrac{21}{L}\right)^{1/x} \left(\dfrac{12}{d}\right)^{1/x} \left(\dfrac{5.125}{b}\right)^{1/x}$
$$= (1.0) \left(\dfrac{21}{49.68 \text{ ft}}\right)^{1/10} \left(\dfrac{12}{39 \text{ in}}\right)^{1/10}$$
$$\times \left(\dfrac{5.125}{8.75 \text{ in}}\right)^{1/10}$$
$$= 0.773$$

The lateral stability factor, C_L, is determined as follows.

The maximum unbraced length, l_u, is the 8 ft purlin spacing. However, in the negative-moment region it is 10.32 ft.

NDS 3.3.3 $l_u = (10.32 \text{ ft}) \left(12 \, \dfrac{\text{in}}{\text{ft}}\right) = 123.8 \text{ in}$

For glulam, $K_{bE} = 0.609$.

Table 3.3.3 $\dfrac{l_u}{d} = \dfrac{123.8 \text{ in}}{39 \text{ in}} = 3.17 < 7$

$l_e = 2.06 l_u = (2.06)(123.8 \text{ in}) = 255.0 \text{ in}$

$R_B = \sqrt{\dfrac{l_e d}{b^2}} = \sqrt{\dfrac{(255.0 \text{ in})(39 \text{ in})}{(8.75 \text{ in})^2}} = 11.4$

$E_y' = E_y C_M = \left(1{,}600{,}000 \, \dfrac{\text{lbf}}{\text{in}^2}\right)(1.0)$
$$= 1{,}600{,}000 \text{ lbf/in}^2$$

$F_{bE} = \dfrac{K_{bE} E_y'}{R_B^2}$
$$= \dfrac{(0.609)\left(1{,}600{,}000 \, \dfrac{\text{lbf}}{\text{in}^2}\right)}{(11.4)^2}$$
$$= 7497.7 \text{ lbf/in}^2$$

$F_b^* = F_{bx} C_D C_M = \left(2400 \, \dfrac{\text{lbf}}{\text{in}^2}\right)(1.15)(1.0)$
$$= 2760.0 \text{ lbf/in}^2$$

$\dfrac{F_{bE}}{F_b^*} = \dfrac{7497.7 \, \dfrac{\text{lbf}}{\text{in}^2}}{2760.0 \, \dfrac{\text{lbf}}{\text{in}^2}} = 2.72$

$\dfrac{1 + \dfrac{F_{bE}}{F_b^*}}{1.9} = \dfrac{1 + 2.72}{1.9} = 1.96$

NDS
Eq. 3.3-6 $C_L = \dfrac{1 + \dfrac{F_{bE}}{F_b^*}}{1.9} - \sqrt{\left(\dfrac{1 + \dfrac{F_{bE}}{F_b^*}}{1.9}\right)^2 - \dfrac{\dfrac{F_{bE}}{F_b^*}}{0.95}}$
$$= 1.96 - \sqrt{(1.96)^2 - \dfrac{2.72}{0.95}}$$
$$= 0.971$$

Since C_V (0.773) is less than C_L (0.971), use C_V.

Check bending stress. The allowable bending stress is

$F_{bx}' = F_{bx} C_D C_M C_V$
$$= \left(2400 \, \dfrac{\text{lbf}}{\text{in}^2}\right)(1.15)(1.0)(0.773)$$
$$= 2133.48 \text{ lbf/in}^2$$

The maximum bending stress is

$f_b = \dfrac{M}{S_x}$
$$= \dfrac{(358.4 \text{ ft-k}) \left(12 \, \dfrac{\text{in}}{\text{ft}}\right) \left(1000 \, \dfrac{\text{lbf}}{\text{k}}\right)}{2218 \text{ in}^3}$$
$$= 1939 \text{ lbf/in}^2$$
$$1939 \text{ lbf/in}^2 < 2133.48 \text{ lbf/in}^2$$
$$f_b < F_{bx}' \quad \text{[OK]}$$

Check the shear stress.

The allowable shear stress is

$$F'_v = F_{vx}C_DC_M$$

$$= \left(190 \; \frac{\text{lbf}}{\text{in}^2}\right)(1.15)(1.0)$$

$$= 218.5 \; \text{lbf/in}^2$$

The maximum shear force and shear stress are $V_{\max} = 40.71$ k (see Fig. 13.5).

$$f_v = \frac{3V}{2A} = \frac{(3)(40.71 \; \text{k})\left(1000 \; \frac{\text{lbf}}{\text{k}}\right)}{(2)(341.2 \; \text{in}^2)}$$

$$= 179 \; \text{lbf/in}^2$$

$$179 \; \text{lbf/in}^2 < 218.5 \; \text{lbf/in}^2$$

$$f_v < F'_v \quad [\text{OK}]$$

For member CD (see Fig. 13.5):

Bending: Member CD has positive moment throughout the entire beam and has continuous lateral support on the compression edge from the decking, assuming that the decking is fastened appropriately (see Fig. 13.6).

Therefore, $C_L = 1.0$.

$C_V = 0.773$ since both members AC and CD are the same glulam size and have the same distance between the regions of zero moment.

The maximum moments and shears are also approximately the same for both members AC and CD. Therefore, bending and shear stress checks are approximately the same for both members. This is a design judgment, and other situations may require the calculation of stresses in each member.

Check deflection.

The largest deflection in the frame rafter occurs in span AB. This deflection is given by

$$E'_x = E_xC_M = \left(1{,}800{,}000 \; \frac{\text{lbf}}{\text{in}^2}\right)(1.0)$$

$$= 1{,}800{,}000 \; \text{lbf/in}^2$$

$$I_x = 43{,}253 \; \text{in}^4$$

App: Beam Formulas

The rafter load was found to be $w = 1158.4 \; \text{lbf/in}^2$.

$$\Delta_{\max} = \left(\frac{0.370}{48}\right)\left(\frac{wL^4}{E'_xI_x}\right)$$

$$= \left(\frac{0.370}{48}\right)\left(\frac{\left(1158.4 \; \frac{\text{lbf}}{\text{ft}}\right)(60 \; \text{ft})^4\left(12 \; \frac{\text{in}}{\text{ft}}\right)^3}{\left(1{,}800{,}000 \; \frac{\text{lbf}}{\text{in}^2}\right)(43{,}253 \; \text{in}^4)}\right)$$

$$= 2.57 \; \text{in}$$

$$\Delta_{\text{allow}} = \frac{L}{180} = \frac{(60 \; \text{ft})\left(12 \; \frac{\text{in}}{\text{ft}}\right)}{180} = 4 \; \text{in}$$

$$4 \; \text{in} > 2.57 \; \text{in}$$

$$\Delta_{\text{allow}} > \Delta_{\max} \quad [\text{OK}]$$

However, camber is also necessary to prevent visible sagging of the frame. Determination of the amount of camber required is beyond the scope of this text (it is often given in prevailing design specifications).

Use glulam $8^3\!/\!_4$ in × 39 in, 24F-V8 douglas fir for frame rafter as a two-span cantilever beam system.

D. Connection design

Between the frame rafter and purlins, the reaction from the purlin to the rafter is

$$R = (\text{purlin load})(\text{purlin span length})\left(\tfrac{1}{2}\right)$$

$$= \left(283.6 \; \frac{\text{lbf}}{\text{ft}}\right)(30 \; \text{ft})\left(\frac{1}{2}\right) = 4254 \; \text{lbf}$$

Using 1 in bolts and 4 in × 4 in × $^1\!/\!_4$ in steel angles, as seen in Fig. 13.6, acting at a 90° angle to the grain of the wood, the bolt design value is

NDS Table 8.3D

$$Z = 1410 \; \text{lbf} \quad (Z_\perp)$$

$$t_m = 3^1\!/\!_8 \; \text{in (purlin)}$$

$$t_s = {}^1\!/\!_4 \; \text{in}$$

For $t_m = 8^3\!/\!_4$ in (frame rafter), use $Z_\perp = 1410$ lbf to be conservative.

The allowable design value is

NDS Table 7.3.1

$$Z' = ZC_DC_MC_tC_gC_\Delta$$

For snow, $C_D = 1.15$.

For dry condition (i.e., moisture content $\leq 16\%$), $C_M = 1.0$.

For normal temperature, $C_t = 1.0$.

NDS
Table 7.3.6C

For group action, $C_g = 1.0$.

NDS 8.5.2–
8.5.6

$C_\Delta = 1.0$

$$Z' = (1410 \text{ lbf})(1.15)(1.0)(1.0)(1.0)(1.0)$$
$$= 1621.5 \text{ lbf/bolt}$$

The purlin reaction is

$$R = \left(283.6 \frac{\text{lbf}}{\text{ft}}\right)(30 \text{ ft})\left(\frac{1}{2}\right)$$
$$= 4254 \text{ lbf}$$

Since the purlins are connected to both sides of each rafter, the number of bolts required in the rafter is

$$N = \frac{2R}{Z'} = \frac{(2)(4254 \text{ lbf})}{1621.5 \dfrac{\text{lbf}}{\text{bolt}}} = 5.25 \text{ bolts}$$

Use six bolts, three in each side of each purlin (see Fig. 13.6).

The total number of bolts required in each purlin is

$$N = \frac{R}{Z'} = \frac{4254 \text{ lbf}}{1621.5 \dfrac{\text{lbf}}{\text{bolt}}} = 2.62 \text{ bolts}$$

Use two bolts at the end of each purlin for a total of four bolts (see Fig. 13.6).

(Note: The load capacity of steel hardware must also be checked, although it is not done in this problem solution.)

E. Design of exterior ridge column (supporting the ridge)

The column is investigated for two load cases.

From Fig. 13.5, the dead load plus snow load for vertical loads is

$$DL + SL = \left(\frac{1}{2}\right)(28.77 \text{ k})(2) = 28.77 \text{ k}$$

(This load is found by noting that each side of the building contributes column load and that exterior columns support one-half the load that corresponding interior columns support.)

The dead load plus wind load for vertical and lateral loads by ratio of DL to (DL + SL) is (see roof deck design for DL and SL) as follows.

$$\begin{aligned}
\frac{\text{vertical}}{\text{load DL}} &= \frac{(28.77 \text{ k})(DL)}{DL + SL} \\[6pt]
&= \frac{(28.77 \text{ k})\left(13.5 \dfrac{\text{lbf}}{\text{ft}^2}\right)}{33.5 \dfrac{\text{lbf}}{\text{ft}^2}} \\[6pt]
&= 11.59 \text{ k} \quad [\text{Fig. 13.7}]
\end{aligned}$$

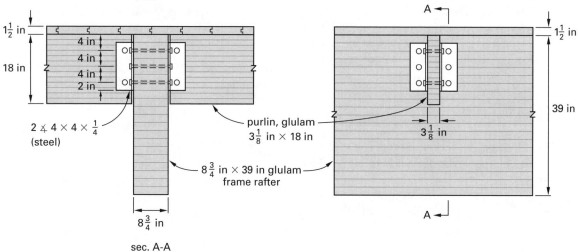

Figure 13.6 Connection Design

lateral
load WL $= $ (wind pressure)(column spacing)

$$= \left(10 \; \frac{\text{lbf}}{\text{ft}^2}\right)(60 \text{ ft})$$

$$= 600 \text{ lbf/ft} \quad [\text{Fig. 13.7}]$$

The calculated column loads are shown in Fig. 13.7.

The column is assumed to be $8^3/4$ in $\times$ 12 in 24F-V10 douglas fir-larch glulam. The wind pressure is assumed to act on the $8^3/4$ in face, which is the wide face of the laminations. It is assumed that the moisture content will not exceed 16%. Normal temperatures (i.e., less than 100°F) apply. Therefore, $C_M = C_t = 1.0$.

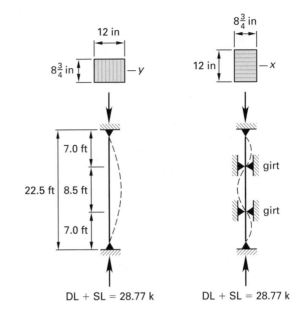

(a) DL + SL

NDS Supp.   The design values are
Table 5A
$F_{bx} = 2400 \text{ lbf/in}^2$
$F_c = 1600 \text{ lbf/in}^2$
$E_x = 1{,}800{,}000 \text{ lbf/in}^2$
$E_y = 1{,}600{,}000 \text{ lbf/in}^2$
$E_{\text{axial}} = 1{,}600{,}000 \text{ lbf/in}^2$

NDS Supp.   The section properties of $8^3/4$ in $\times$ 12 in glu-
Table 1C   lam are
$A = 105 \text{ in}^2$
$S_x = 210 \text{ in}^3$
$S_y = 153.1 \text{ in}^3$
$I_x = 1260 \text{ in}^4$
$I_y = 669.9 \text{ in}^4$

For the dead load + snow load case (see Fig. 13.7(a)), the unbraced length with respect to the x-axis is different than the unbraced length with respect to the y-axis.

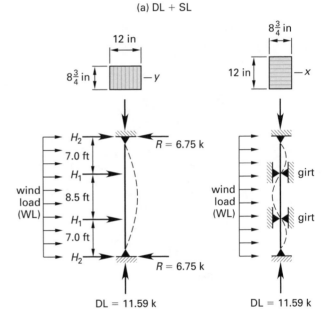

WL = 0.6 k/ft
$H_1 = (0.6 \text{ ft})(7 \text{ ft} + 8.5 \text{ ft})\left(\frac{1}{2}\right) = 4.65 \text{ k}$
$H_2 = \left(0.6 \; \frac{\text{k}}{\text{ft}}\right)(7 \text{ ft})\left(\frac{1}{2}\right) = 2.1 \text{ k}$

(b) DL + WL
(wind load distributed to four
girts attached to the column)

Figure 13.7 Column Loads for Both Load Cases

NDS 3.7   Determine the column stability factor, C_P,
and Eq. 3.7-1   about the x-axis (see Fig. 13.7).

$$\left(\frac{l_e}{d}\right)_x = \left(\frac{K_e l}{d}\right)_x$$

$$= \frac{(1.0)(22.5 \text{ ft})\left(12 \; \frac{\text{in}}{\text{ft}}\right)}{12 \text{ in}}$$

$$= 22.5$$

$$\begin{aligned}
E'_x &= E_x C_M C_t \\
&= \left(1{,}800{,}000 \; \frac{\text{lbf}}{\text{in}^2}\right)(1.0)(1.0) \\
&= 1{,}800{,}000 \text{ lbf/in}^2
\end{aligned}$$

For glulam, $K_{cE} = 0.418$ and $c = 0.9$.

$$F_{cEx} = \frac{K_{cE}E'_x}{\left(\dfrac{l_e}{d}\right)^2_x}$$

$$= \frac{(0.418)\left(1{,}800{,}000\ \dfrac{\text{lbf}}{\text{in}^2}\right)}{(22.5)^2}$$

$$= 1486\ \text{lbf/in}^2$$

For snow, $C_D = 1.15$.

$$F_c^* = F_c C_D C_M C_t$$

$$= \left(1600\ \frac{\text{lbf}}{\text{in}^2}\right)(1.15)(1.0)(1.0)$$

$$= 1840\ \text{lbf/in}^2$$

$$\frac{F_{cEx}}{F_c^*} = \frac{1486\ \dfrac{\text{lbf}}{\text{in}^2}}{1840\ \dfrac{\text{lbf}}{\text{in}^2}} = 0.808$$

$$\frac{1 + \dfrac{F_{cEx}}{F_c^*}}{2c} = \frac{1 + 0.808}{(2)(0.9)} = 1.0$$

$$C_P = \frac{1 + \dfrac{F_{cEx}}{F_c^*}}{2c} - \sqrt{\left(\frac{1 + \dfrac{F_{cEx}}{F_c^*}}{2c}\right)^2 - \frac{\dfrac{F_{cEx}}{F_c^*}}{c}}$$

$$= 1.0 - \sqrt{(1.0)^2 - \frac{0.808}{0.9}}$$

$$= 0.680$$

Determine the column stability factor, C_P, about the y-axis (see Fig. 13.7).

$$\left(\frac{l_e}{d}\right)_y = \left(\frac{K_e l}{d}\right)_y$$

$$= \frac{(1.0)(8.5\ \text{ft})\left(12\ \dfrac{\text{in}}{\text{ft}}\right)}{8.75\ \text{in}}$$

$$= 11.7$$

$$E'_y = E_y C_M C_t$$

$$= \left(1{,}600{,}000\ \frac{\text{lbf}}{\text{in}^2}\right)(1.0)(1.0)$$

$$= 1{,}600{,}000\ \text{lbf/in}^2$$

$$F_{cEy} = \frac{K_{cE}E'_y}{\left(\dfrac{l_e}{d}\right)^2_y}$$

$$= \frac{(0.418)\left(1{,}600{,}000\ \dfrac{\text{lbf}}{\text{in}^2}\right)}{(11.7)^2}$$

$$= 4886\ \text{lbf/in}^2$$

$$\frac{F_{cEy}}{F_c^*} = \frac{4886\ \dfrac{\text{lbf}}{\text{in}^2}}{1840\ \dfrac{\text{lbf}}{\text{in}^2}} = 2.66$$

$$\frac{1 + \dfrac{F_{cEy}}{F_c^*}}{2c} = \frac{1 + 2.66}{(2)(0.9)} = 2.03$$

$$C_P = \frac{1 + \dfrac{F_{cEy}}{F_c^*}}{2c} - \sqrt{\left(\frac{1 + \dfrac{F_{cEy}}{F_c^*}}{2c}\right)^2 - \frac{\dfrac{F_{cEy}}{F_c^*}}{c}}$$

$$= 2.03 - \sqrt{(2.03)^2 - \frac{2.66}{0.9}}$$

$$= 0.950$$

The x-axis produces the smaller value for the column stability factor, C_P. The allowable compressive stress is

$$F'_c = F_c C_D C_M C_t C_P$$

$$= \left(1600\ \frac{\text{lbf}}{\text{in}^2}\right)(1.15)(1.0)(1.0)(0.680)$$

$$= 1251.2\ \text{lbf/in}^2$$

The allowable column load is

$$P_{\text{allow}} = F'_c A = \left(1251.2\ \frac{\text{lbf}}{\text{in}^2}\right)(105\ \text{in}^2)$$

$$= 131{,}376\ \text{lbf}$$

$$131{,}376\ \text{lbf} > 28{,}770\ \text{lbf}\ \left[\text{see Fig. 13.5}\right]$$

$$P_{\text{allow}} > \text{DL} + \text{SL}\quad [\text{OK}]$$

For dead load + wind load case (see Fig. 13.7(b)):

The actual axial stress from dead load is

$$f_c = \frac{P}{A} = \frac{11{,}590\ \text{lbf}}{105\ \text{in}^2} = 110.4\ \text{lbf/in}^2$$

As calculated previously from the DL + SL case, the column stability factor, C_P, about the x-axis is 0.680, and the allowable compressive stress of the column is 1251.2 lbf/in^2. The axial stress ratio is

$$\frac{f_c}{F'_c} = \frac{110.4 \; \frac{lbf}{in^2}}{1251.2 \; \frac{lbf}{in^2}} = 0.088$$

Bending from wind load (see Fig. 13.7):

Based on the distribution of wind load, on tributary areas, from the siding to four girts attached to the column, the bending moment about the x-axis is

$$M_x = (6750 \; lbf - 2100 \; lbf)(7 \; ft)$$
$$= 32{,}550 \; \text{ft-lbf} \quad [\text{see Fig. 13.7(b)}]$$

The actual bending stress is

$$f_{bx} = \frac{M_x}{S_x} = \frac{(32{,}550 \; \text{ft-lbf})\left(12 \; \frac{in}{ft}\right)}{210 \; in^3}$$
$$= 1860 \; lbf/in^2$$

NDS Table 2.3.1 The allowable bending stress is the smaller of $F'_{bx} = F_{bx}C_D C_M C_t C_L$ and $F'_{bx} = F_{bx}C_D C_M C_t C_V$.

NDS 3.3.3; Eq. 3.3-6 Determine the beam stability factor, C_L. Lateral-torsional buckling can occur in the plane of the y-axis. The unbraced length, l_u, is 8.5 ft (102 in).

$$\frac{l_u}{d} = \frac{102 \; in}{12 \; in} = 8.5 > 7$$

NDS Table 3.3.3
$$l_e = 1.63 l_u + 3d$$
$$= (1.63)(102 \; in) + (3)(12 \; in)$$
$$= 202.26 \; in$$

$$R_B = \sqrt{\frac{l_e d}{b^2}} = \sqrt{\frac{(202.26 \; in)(12 \; in)}{(8.75 \; in)^2}} = 5.63$$

$$E'_y = E_y C_M C_t$$
$$= \left(1{,}600{,}000 \; \frac{lbf}{in^2}\right)(1.0)(1.0)$$
$$= 1{,}600{,}000 \; lbf/in^2$$

For glulam, $K_{bE} = 0.609$.

$$F_{bE} = \frac{K_{bE} E'_y}{R_B^2}$$

$$= \frac{(0.069)\left(1{,}600{,}000 \; \frac{lbf}{in^2}\right)}{(5.63)^2}$$

$$= 30{,}741 \; lbf/in^2$$

For wind/earthquake loads, $C_D = 1.6$.

$$F_b^* = F_{bx}C_D C_M C_t$$
$$= \left(2400 \; \frac{lbf}{in^2}\right)(1.6)(1.0)(1.0)$$
$$= 3840 \; lbf/in^2$$

$$\frac{F_{bE}}{F_b^*} = \frac{30{,}741 \; \frac{lbf}{in^2}}{3840 \; \frac{lbf}{in^2}} = 8.0$$

$$\frac{1 + \dfrac{F_{bE}}{F_b^*}}{1.9} = \frac{1 + 8.0}{1.9} = 4.74$$

$$C_L = \frac{1 + \dfrac{F_{bE}}{F_b^*}}{1.9} - \sqrt{\left(\frac{1 + \dfrac{F_{bE}}{F_b^*}}{1.9}\right)^2 - \frac{\dfrac{F_{bE}}{F_b^*}}{0.95}}$$

$$= 4.74 - \sqrt{(4.74)^2 - \frac{8.0}{0.95}}$$

$$= 0.99$$

NDS 5.3.2; Eq. 5.3-1; NDS Supp. Table 5A Determine the volume factor, C_V.

$$C_V = K_L \left(\frac{21}{L}\right)^{1/x} \left(\frac{12}{d}\right)^{1/x} \left(\frac{5.125}{b}\right)^{1/x}$$
$$\leq 1.0$$

NDS Table 5.3.2 For two equal concentrated loads, $K_L = 0.96$.

For all other than southern pine, $x = 10$.

$$L = 22.5 \; ft$$

$$C_V = (0.96)\left(\frac{21}{22.5 \; ft}\right)^{1/10}\left(\frac{12}{12 \; in}\right)^{1/10}$$
$$\times \left(\frac{5.125}{8.75 \; in}\right)^{1/10}$$

$$= 0.90$$
$$0.90 < 0.99$$
$$C_V < C_L$$

Therefore, use C_V to determine the allowable bending stress.

$$F'_{bx} = F_{bx}C_DC_MC_tC_V$$
$$= \left(2400 \ \frac{\text{lbf}}{\text{in}^2}\right)(1.6)(1.0)(1.0)(0.90)$$
$$= 3456 \ \text{lbf/in}^2$$

NDS Eq. 3.9-3 Combined compression and bending stresses are

$$\left(\frac{f_c}{F'_c}\right)^2 + \frac{f_{bx}}{F'_{bx}\left(1 - \frac{f_c}{F_{cEx}}\right)} \le 1.0$$

From the dead load + snow load analysis, $F_{cEx} = 1486 \ \text{lbf/in}^2$.

$$\left(\frac{110.4 \ \frac{\text{lbf}}{\text{in}^2}}{1251.2 \ \frac{\text{lbf}}{\text{in}^2}}\right)^2 + \frac{1860 \ \frac{\text{lbf}}{\text{in}^2}}{\left(3456 \ \frac{\text{lbf}}{\text{in}^2}\right)\left(1 - \frac{110.4 \ \frac{\text{lbf}}{\text{in}^2}}{1486 \ \frac{\text{lbf}}{\text{in}^2}}\right)}$$

$$= 0.589 < 1.0 \quad [\text{OK}]$$

Therefore, use $8^3/_4$ in $\times$ 12 in 24F-V3 douglas fir column.

Practice Problem 4: Bolted Splice Connection

The splice connection shown uses a row of $^3/_4$ in bolts. The lumber is no. 2 grade eastern white pine. The load is caused by dead load + snow load. It is assumed that $C_M = 1.0$ and $C_t = 1.0$.

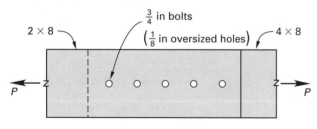

elevation view

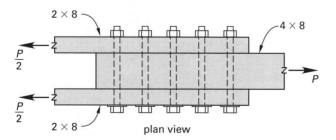

plan view

Determine the tension capacity of the splice and the number of $^3/_4$ in bolts required. Assume bolt spacing and edge distance requirements are met for full design values.

Solution:

Reference

NDS Table 2.3.1

A. Determine lumber capacity (two 2 × 8s control).

$$F'_t = F_tC_DC_MC_tC_F$$

NDS Supp. Table 4A and Table 4A Adj. Factors

For no. 2 eastern white pine, $F_t = 275 \ \text{lbf/in}^2$.

$C_M = 1.0$

$C_t = 1.0$

For 2 × 8 for F_t, $C_F = 1.2$.

NDS Table 2.3.2

For snow, $C_D = 1.15$.

Therefore, the allowable stress is

$$F'_t = \left(275 \ \frac{\text{lbf}}{\text{in}^2}\right)(1.15)(1.0)(1.0)(1.2)$$
$$= 379.5 \ \text{lbf/in}^2$$

The lumber capacity based on gross area is

$$P_{\text{allow}} = F'_tA_g$$
$$= \left(379.5 \ \frac{\text{lbf}}{\text{in}^2}\right)(1.5 \ \text{in})(7.25 \ \text{in})$$
$$\times (2 \ \text{members})$$
$$= 8254 \ \text{lbf}$$

The lumber capacity based on net area is

$$P_{\text{allow}} = F'_tA_n$$
$$= \left(379.5 \ \frac{\text{lbf}}{\text{in}^2}\right)(1.5 \ \text{in})$$
$$\times \left(7.25 \ \text{in} - (0.75 \ \text{in} + 0.125 \ \text{in})\right)$$
$$\times (2 \ \text{members})$$
$$= 7258 \ \text{lbf} \quad [\text{controls}]$$

B. Determine the number of $^3/_4$ in bolts required.

NDS Table 7.3.1

The allowable bolt design value is

$$Z' = ZC_DC_MC_tC_gC_\Delta$$

NDS
Table 8.3A

For lateral design value with all wood members loaded parallel to grain, $Z = Z_\| = 1990$ lbf for $t_m = 3\frac{1}{2}$ in, $t_s = 1\frac{1}{2}$ in, eastern softwoods, and $\frac{3}{4}$ in bolt.

$$C_D = 1.0$$

$$C_M = 1.0$$

$$C_t = 1.0$$

NDS
Table 7.3.6A

By interpolation (or use NDS Eq. 7.3-1 for more accuracy), the group action factor is

$C_g \approx 0.90$ for
$$A_s = (2)(1.5 \text{ in})(7.25 \text{ in}) = 21.75 \text{ in}^2$$
$$A_m = (3.5 \text{ in})(7.25 \text{ in}) = 25.375 \text{ in}^2$$
$$\frac{A_s}{A_m} = \frac{21.75 \text{ in}^2}{25.375 \text{ in}^2} = 0.857$$
Assume five inline bolts.

NDS 8.5.2–
8.5.6

The geometry factor is $C_\Delta = 1.0$ if
$$\begin{aligned} \text{end distance} &\geq 7D = (7)(\tfrac{3}{4} \text{ in}) \\ &= 5.25 \text{ in} \\ \text{bolt spacing} &\geq 4D = (4)(\tfrac{3}{4} \text{ in}) \\ &= 3 \text{ in} \\ \text{edge distance} &\geq 1.5D = (1.5)(\tfrac{3}{4} \text{ in}) \\ &= 1.125 \text{ in} \end{aligned}$$

NDS
Table 7.3.1

$$\begin{aligned} Z' &= (1990 \text{ lbf})(1.15)(1.0)(1.0)(0.90)(1.0) \\ &= 2060 \text{ lbf per bolt} \end{aligned}$$

The total number of $\frac{3}{4}$ in bolts required is

$$\begin{aligned} N &= \frac{\text{lumber capacity}}{\text{allowable bolt design value}} \\ &= \frac{7258 \text{ lbf}}{2060 \dfrac{\text{lbf}}{\text{bolt}}} = 3.52 \text{ bolts} \end{aligned}$$

Use four $\frac{3}{4}$ in bolts for a tension capacity of 7258 lbf. Follow necessary requirements for bolt edge distance, end distance, and spacing. There is no need to recalculate for four bolts since C_g gets larger.

Practice Problem 5:
Bolted Splice Connection with Metal Side Plates

Use the same information as in Prob. 4 except with $\frac{1}{4}$ in $\times$ 2 in ASTM steel side plates instead of 2 $\times$ 8 wood side plates.

Determine the number of $\frac{3}{4}$ in bolts required based on the tension capacity of the lumber.

Solution:

Reference

NDS Table
7.3.1, 8.3B

The allowable bolt design value is

$$Z' = ZC_DC_MC_tC_gC_\Delta$$

$$Z = Z_\| = 2660 \text{ lbf}$$

$$C_D = 1.15$$

$$C_M = 1.0$$

$$C_t = 1.0$$

The lumber capacity of no. 2 4 $\times$ 8 eastern white pine is

NDS
Table 4A

$$F_t = 275 \text{ lbf/in}^2$$

$$C_F = 1.2 \text{ for } 4 \times 8$$

$$\begin{aligned} F_t' &= F_tC_DC_MC_tC_F \\ &= \left(275 \frac{\text{lbf}}{\text{in}^2}\right)(1.15)(1.0)(1.0)(1.2) \\ &= 380 \text{ lbf/in}^2 \end{aligned}$$

$$\begin{aligned} P_{\text{allow}} &= F_t'A_n \\ &= \left(380 \frac{\text{lbf}}{\text{in}^2}\right)(3.5 \text{ in}) \\ &\quad \times \left(7.25 \text{ in} - (0.75 \text{ in} + 0.125 \text{ in})\right) \\ &= 8479 \text{ lbf} \end{aligned}$$

NDS
Table 7.3.6C

The group action factor is $C_g \approx 0.97$ for

$$\begin{aligned} A_s &= (2)(0.25 \text{ in})(2 \text{ in}) = 1.0 \text{ in}^2 \\ A_m &= (3.5 \text{ in})(7.25 \text{ in}) \\ &= 25.375 \text{ in}^2 \text{ for } 4 \times 8 \\ \frac{A_m}{A_s} &= \frac{25.375 \text{ in}^2}{1.0 \text{ in}^2} = 25.375 \end{aligned}$$
Assume four bolts in a row.

(Or use NDS Eq. 7.3-1 for more accuracy.)

NDS 8.5.2–
8.5.6

The geometry factor is $C_\Delta = 1.0$ if the necessary bolt edge distance, end distance, and spacing requirements are met.

NDS
Table 7.3.1

$$\begin{aligned} Z' &= (2660 \text{ lbf})(1.15)(1.0)(1.0)(0.97)(1.0) \\ &= 2967 \text{ lbf per bolt} \end{aligned}$$

The number of $\frac{3}{4}$ in bolts required is

$$N = \frac{8479 \text{ lbf}}{2967 \dfrac{\text{lbf}}{\text{bolt}}} = 2.86 \text{ bolts}$$

Therefore, use three ³/₄ in bolts. There is no need to recalculate with three bolts since C_g gets larger.

Practice Problem 6:
Splice Connection with Spikes

Revise Prob. 4 such that the connectors used are 60d common wire spikes instead of ³/₄ in bolts.

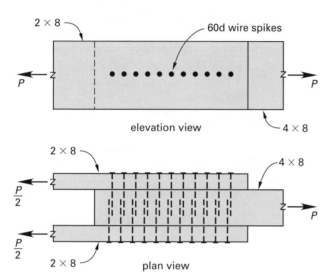

elevation view

plan view

Determine the number of 60d common wire spikes required for Prob. 4. The two 2×8 lumber capacities based on gross area were 8254 lbf.

Solution:

Reference

Determine the dimensions and design value for 60d pennyweight wire spikes.

NDS
Table 12.3D

For eastern white pine, $t_s = 1.5$ in, and a 60d spike,

$L = 6$ in

$D = 0.283$ in

$Z = 166$ lbf

The allowable design value for a 60d spike is

NDS
Table 7.3.1

$Z' = ZC_D C_M C_t C_d C_{eg} C_{di} C_{tn}$

$C_{eg}, C_{di}, C_{tn} = $ not applicable

For snow, $C_D = 1.15$.

For normal temperatures, $C_t = 1.0$.

For dry in-service conditions, $C_M = 1.0$.

NDS 12.3.4   The penetration depth factor is C_d.

The actual penetration into the main 4×8 member is
$p = 6$ in $- 1.5$ in $= 4.5$ in

$$C_d = \frac{p}{12D} \leq 1.0$$

$$C_d = \frac{4.5 \text{ in}}{(12)(0.283 \text{ in})} = 1.32$$

Therefore, $C_d = 1.0$.

$$Z' = ZC_D C_M C_t C_d$$

$$Z' = (166 \text{ lbf})(1.15)(1.0)(1.0)(1.0) = 191 \text{ lbf}$$

The number of spikes required for one-half of 8254 lbf is

$$N = \frac{\dfrac{8254 \text{ lbf}}{2 \text{ side plates}}}{191 \dfrac{\text{lbf}}{\text{spike}}} = 21.6 \text{ spikes}$$

Use 22 60d spikes on each side plate.

Practice Problem 7:
Truss End Analysis

The end of a no. 1 eastern white pine truss end is shown.

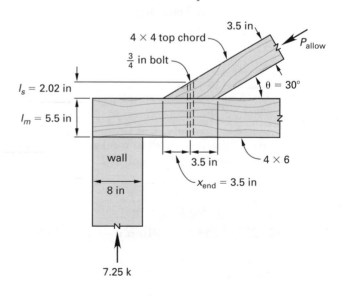

Determine the compressive load capacity of the 4×4 top chord as based on bolt capacity.

Solution:

Reference

NDS 8.2.4, Table 8.5.4 8.5.4.3;

Determine the geometry factor, C_Δ.

For the given end distance and loading at an angle, the minimum shear area for the full design value will be equivalent to $4Dl_s$ for a parallel member in compression.

$$4Dl_s = (4)(0.75 \text{ in})(2.02 \text{ in}) = 6.06 \text{ in}^2$$

For the reduced design value, the shear area is

$$2Dl_s = (2)(0.75 \text{ in})(2.02 \text{ in}) = 3.03 \text{ in}^2$$

The actual equivalent shear area is (see illustration and NDS Fig. 8B)

$$\tfrac{1}{2}x_{\text{end}}l_s = \left(\frac{1}{2}\right)(3.5 \text{ in})(2.02 \text{ in})$$

$$= 3.5 \text{ in}^2 > 3.03 \text{ in}^2 \quad [\text{OK}]$$

$$C_\Delta = \frac{\text{actual shear area}}{\text{shear area for full design}}$$

$$= \frac{3.5 \text{ in}^2}{6.06 \text{ in}^2} = 0.578$$

Determine the allowable bolt design value. For a ³⁄₄ in bolt with $t_m = l_m = 5.5$ in, $t_s = l_s = 2.02$ in, and eastern white pine lumber,

NDS Table 8.2A

$$Z' = Z_\| C_D C_M C_g C_\Delta$$

$$Z_\| = 1089 \text{ lbf} \quad [\text{by interpolation}]$$

$$C_D = 1.15 \text{ for snow}$$

$$C_M = 1.0 \text{ for dry in-service conditions}$$

$$C_g = 1.0 \text{ for one bolt}$$

$$C_\Delta = 0.578$$

$$Z' = (1089 \text{ lbf})(1.15)(1.0)(1.0)(0.578)$$

$$= 724 \text{ lbf} \quad \begin{bmatrix} \text{load capacity acting} \\ \text{perpendicular to bolt axis} \end{bmatrix}$$

The maximum allowed compression in top chord based on bolt capacity is

$$P_{\text{allow}} = \frac{Z'}{\cos\theta} = \frac{724 \text{ lbf}}{\cos 30°} = 836 \text{ lbf}$$

Appendix:
Beam Formulas

1. Concentrated Load, P, in Simple Beams

A. At Any Point in Span

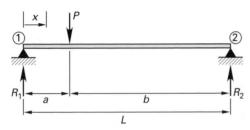

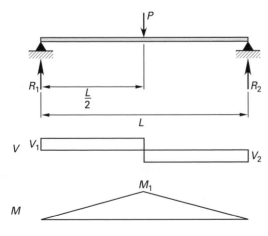

$$R_1 = V_1 = \frac{Pb}{L}$$

$$R_2 = -V_2 = \frac{Pa}{L}$$

$$M_1 = M_{max} = \frac{Pab}{L}$$

$$\Delta_a \text{ (at point of load)} = \frac{Pa^2b^2}{3EIL}$$

$$\Delta_x \text{ (when } x < a) = \left(\frac{Pbx}{6EIL}\right)(L^2 - b^2 - x^2)$$

$$\Delta_{max} = \frac{Pab(L+b)\sqrt{3a(L+b)}}{27EIL} \text{ at } \sqrt{\frac{a(L+b)}{3}} \text{ from } ①$$

B. At Midspan

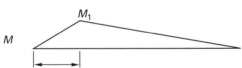

$$R_1 = V_1 = \frac{P}{2}$$

$$R_2 = -V_2 = \frac{P}{2}$$

$$M_1 = M_{max} = \frac{PL}{4}$$

$$\Delta_{max} = \frac{PL^3}{48EI} \text{ at midspan}$$

C. Beam Fixed at Both Ends

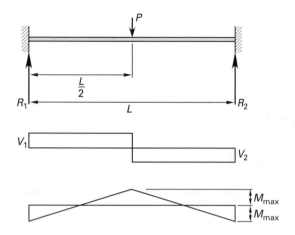

$$R_1 = R_2 = V_1 = -V_2 = \frac{P}{2}$$

$$M_{max} \text{ (at center and ends)} = \pm \frac{PL}{8}$$

$$\Delta_{max} \text{ (at center)} = \frac{PL^3}{192EI}$$

D. Cantilever Beam

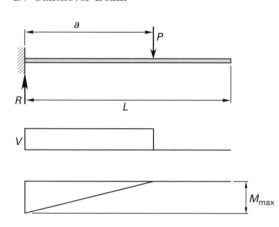

$$R = V = P$$

$$M_{max} \text{ (at fixed end)} = -Pa$$

$$\Delta_{max} \text{ (at free end)} = \left(\frac{Pa^2}{6EI}\right)(3L - a)$$

2. Two Equal Spans: Concentrated Load

A. Concentrated Load at Center of One Span

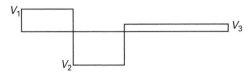

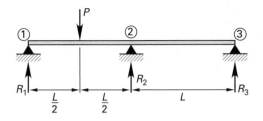

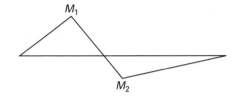

$$R_1 = V_1 = \tfrac{13}{32}P$$

$$R_2 = |V_2| + |V_3| = \tfrac{11}{16}P$$

$$R_3 = -V_3 = -\tfrac{3}{32}P$$

$$V_2 = -\tfrac{19}{32}P$$

$$M_1 = \tfrac{13}{64}PL$$

$$M_2 = -\tfrac{3}{32}PL$$

$$\Delta_{\max} = \frac{0.015PL^3}{EI} \quad \text{at } 0.48L \text{ from } \text{①}$$

B. Concentrated Load at Any Point of One Span

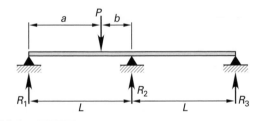

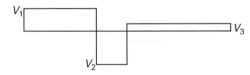

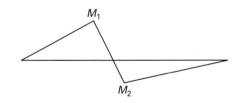

$$R_1 = V_1 = \left(\frac{Pb}{4L^3}\right)\left(4L^2 - a(L+a)\right)$$

$$R_2 = |V_2| + |V_3| = \left(\frac{Pa}{2L^3}\right)\left(2L^2 + b(L+a)\right)$$

$$R_3 = -V_3 = \left(\frac{-Pab}{4L^3}\right)(L+a)$$

$$V_2 = \left(\frac{-Pa}{4L^3}\right)\left(4L^2 + b(L+a)\right)$$

$$M_1 = M_{\max} = \left(\frac{Pab}{4L^3}\right)\left(4L^2 - a(L+a)\right)$$

$$M_2 = \left(\frac{-Pab}{4L^2}\right)(L+a)$$

C. Continuous Beam of Two Equal Spans: Equal Concentrated Loads, P, at Center of Each Span

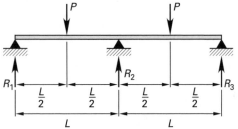

reactions: $R_1 = R_3 = \frac{5}{16}P$

$$R_2 = 1.375\ P$$

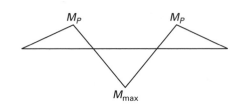

shear forces: $V_1 = -V_3 = \frac{5}{16}P$

$$V_2 = \pm\frac{11}{16}P$$

bending moments:
$$M_{\max} = -\frac{6}{32}PL \text{ at } R_2$$

$$M_p = \frac{5}{32}PL \text{ at points of loads}$$

D. Continuous Beam of Two Equal Spans: Concentrated Loads, P, at Third Points of Each Span

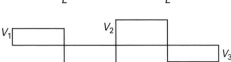

reactions: $R_1 = R_3 = \frac{2}{3}P$

$$R_2 = \frac{8}{3}P$$

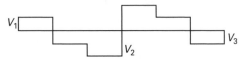

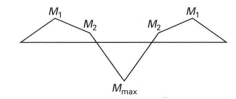

shear forces: $V_1 = -V_3 = \frac{2}{3}P$

$$V_2 = \pm\frac{4}{3}P$$

bending moments:
$$M_{\max} = -\frac{1}{3}PL \text{ at } R_2$$

$$M_1 = \frac{2}{9}PL$$

$$M_2 = \frac{1}{9}PL$$

3. Uniformly Distributed Loads: One Span Beam

A. Simple Beam

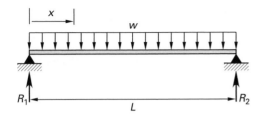

$$R_1 = R_2 = \frac{wL}{2}$$

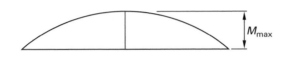

$$V_1 = -V_2 = R_1 = R_2$$

$$M_{\max} = \frac{wL^2}{8}$$

$$\Delta_x = \left(\frac{wx}{24EI}\right)(L^3 - 2Lx^2 + x^3)$$

$$\Delta_{\max} \text{ (at center)} = \frac{5wL^4}{384EI}$$

B. Beam Fixed at Both Ends

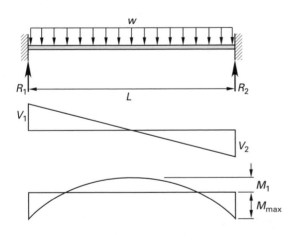

$$R_1 = R_2 = \frac{wL}{2}$$

$$V_1 = -V_2 = R_1 = R_2$$

$$M_1 \text{ (at center)} = \frac{wL^2}{24}$$

$$M_{\max} \text{ (at ends)} = \frac{-wL^2}{12}$$

$$\Delta_{\max} \text{ (at center)} = \frac{wL^4}{384EI}$$

C. Cantilever Beam

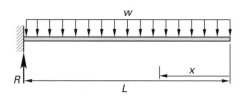

$$R = V = wL$$

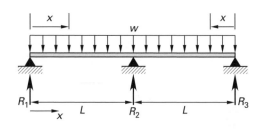

$$M_{\max} \text{ (at fixed end)} = \frac{-wL^2}{2}$$

$$\Delta_x = \left(\frac{w}{24EI}\right)(x^4 - 4L^3x + 3L^4)$$

$$\Delta_{\max} \text{ (at free end)} = \frac{wL^4}{8EI}$$

4. Two Equal Spans: Uniform Load

A. Uniform Load on Both Spans, w

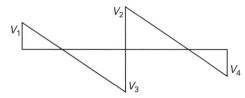

$$R_1 = V_1 = \tfrac{3}{8}wL$$

$$R_2 = |V_2| + |V_3| = \tfrac{10}{8}wL$$

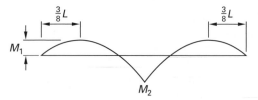

$$V_2 = -V_3 = \tfrac{5}{8}wL$$

$$R_3 = -V_4 = \tfrac{3}{8}wL$$

$$M_1 = \tfrac{9}{128}wL^2$$

$$M_2 = -\tfrac{1}{8}wL^2$$

$$\Delta_x = \left(\frac{w}{48EI}\right)(L^3x - 3Lx^3 + 2x^4),\ 0 \le x \le L$$

$$\Delta_{\max} = \left(\frac{1}{185}\right)\left(\frac{wL^4}{EI}\right) \text{ at } 0.42L \text{ from } R_1 \text{ or } R_3$$

B. Uniform Load on One Span

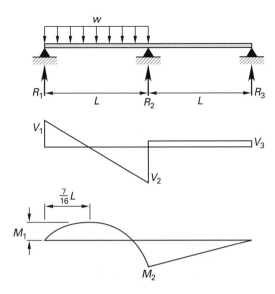

$$R_1 = V_1 = \tfrac{7}{16}wL$$

$$R_2 = |V_2| + |V_3| = \tfrac{5}{8}wL$$

$$R_3 = -V_3 = -\tfrac{1}{16}wL$$

$$V_2 = \tfrac{-9}{16}wL$$

$$M_1 = \tfrac{49}{512}wL^2$$

$$M_2 = -\tfrac{1}{16}wL^2$$

$$\Delta_{\max} = \left(\frac{1}{109EI}\right)wL^4 \text{ at } 0.472L \text{ from } R_1$$

5. Beam Overhang over One Support

A. Beam Overhang: Uniformly Distributed Load

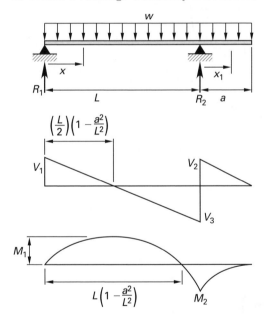

$$R_1 = V_1 = \left(\frac{w}{2L}\right)\left(L^2 - a^2\right)$$

$$R_2 = |V_2| + |V_3| = \left(\frac{w}{2L}\right)(L+a)^2$$

$$V_2 = wa$$

$$V_3 = \left(\frac{-w}{2L}\right)\left(L^2 + a^2\right)$$

$$M_1 = \left(\frac{w}{8L^2}\right)(L+a)^2(L-a)^2$$

$$M_2 = -\frac{wa^2}{2}$$

$$\Delta_{x_1} \text{ (along overhang)} = \left(\frac{wx_1}{2EI}\right)\left(4a^2L - L^3 + 6a^2x_1\right.$$
$$\left. - 4ax_1^2 + x_1^3\right)$$

$$\Delta_{\text{overhang}} = \left(\frac{wa}{24EI}\right)\left(4a^2L - L^3 + 3a^3\right)$$

$$\Delta_x \text{ (between supports)}$$
$$= \left(\frac{wx}{24EIL}\right)\left(L^4 - 2L^2x^2 + Lx^3\right.$$
$$\left. - 2a^2L^2 + 2a^2x^2\right)$$

B. Beam Overhang: Uniform Load on Overhang

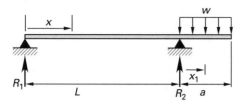

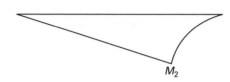

$$R_1 = \frac{-wa^2}{2L}$$

$$V_1 = R_1$$

$$R_2 = |V_1| + |V_2| = \left(\frac{wa}{2L}\right)(2L + a)$$

$$V_2 = wa$$

$$M_2 = -\frac{wa^2}{2}$$

$$\Delta_x \text{ (between supports)} = \left(\frac{wa^2 x}{12EIL}\right)(L^2 - x^2)$$

$$\Delta_{\max} \left(\text{between supports at } x = \frac{L}{\sqrt{3}}\right)$$

$$= (0.03208)\left(\frac{wa^2 L^2}{EI}\right)$$

$$\Delta_{x_1} \text{ (for overhang)}$$

$$= \left(\frac{wx_1}{24EI}\right)(4a^2 L + 6a^2 x_1 - 4ax_1^2 + x_1^3)$$

$$\Delta_{\text{overhang}} = \left(\frac{wa^3}{24EI}\right)(4L + 3a)$$

C. Beam Overhang: Concentrated Load Between Supports

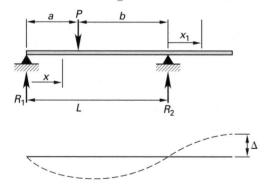

$$R_1 = \frac{Pb}{L}$$

$$R_2 = \frac{Pa}{L}$$

$$V, M = \text{case 1(A)}$$

$$\Delta_{\max} \left(\text{at } x = \sqrt{\frac{a(a + 2b)}{3}} \text{ when } a > b\right)$$

$$= \frac{Pab(a + 2b)\sqrt{3a(a + 2b)}}{27EIL}$$

$$\Delta_a \text{ (at point of load)} = \frac{Pa^2 b^2}{3EIL}$$

$$\Delta_{x_1} = \left(\frac{Pabx_1}{6EIL}\right)(L + a)$$

D. Beam Overhang: Concentrated Load at Overhang

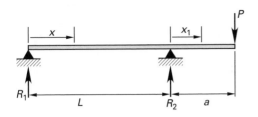

$$R_1 = -\frac{Pa}{L}$$

$$R_2 = \left(\frac{P}{L}\right)(L + a)$$

$$V_1 = R_1 = \frac{-Pa}{L}$$
$$V_2 = P$$

$$M = -Pa$$

$$\Delta_x \text{ (between supports)} = \left(\frac{Pax}{6EIL}\right)(L^2 - x^2)$$

$$\Delta_{\max}\left(\text{between supports at } x = \frac{L}{\sqrt{3}}\right)$$
$$= (0.06415)\left(\frac{PaL^2}{EI}\right)$$

$$\Delta_{x_1} \text{ (in overhang)} = \left(\frac{Px_1}{6EI}\right)(2aL + 3ax_1 - x_1^2)$$

$$\Delta_{\text{overhang}} = \left(\frac{Pa^2}{3EI}\right)(L + a)$$

E. Beam Overhang, One Support: Uniformly Distributed Load Between Supports

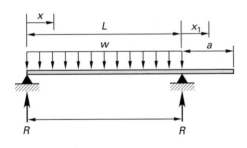

$$R = \frac{WL}{2}$$

$$V = \frac{WL}{2}$$

$$M_{\max} \text{ (at center)} = \frac{WL^2}{8}$$

$$\Delta_{\max} \text{ (at center)} = \frac{5WL^4}{384EI}$$

$$\Delta_x = \left(\frac{wx}{24EI}\right)(L^3 - 2Lx^2 + x^3)$$

$$\Delta_{x_1} = \Delta_{\text{overhang}} = \frac{WL^3 x_1}{24EI} \text{ or } \frac{WL^3 a}{24EI}$$

6. Cantilevered Beam: Two Equal Spans

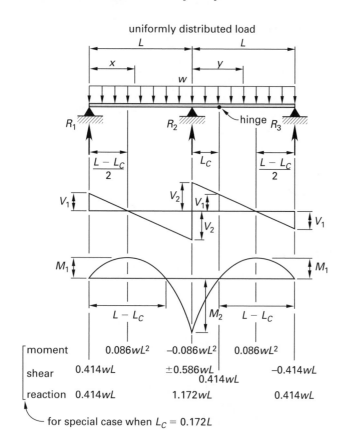

General formulas:

$$R_1 = R_3 = \left(\frac{w}{2}\right)(L - L_C)$$

$$R_2 = w(L + L_C)$$

$$V_1 = \pm\left(\frac{w}{2}\right)(L - L_C) \qquad V_x = \left(\frac{w}{2}\right)(L - L_C - 2x)$$

$$V_2 = \pm\left(\frac{w}{2}\right)(L + L_C) \qquad V_y = \left(\frac{w}{2}\right)(L + L_C - 2y)$$

$$M_1 = \left(\frac{w}{8}\right)(L - L_C)^2$$

$$M_2 = -\frac{wLL_C}{2}$$

$$M_x = \left(\frac{wx}{2}\right)(L - L_C - x)$$

$$M_y = \left(\frac{w}{2}\right)(y - L_C)(L - y)$$

moment	$0.086wL^2$	$-0.086wL^2$	$0.086wL^2$
shear	$0.414wL$	$\pm0.586wL$	$-0.414wL$
		$0.414wL$	
reaction	$0.414wL$	$1.172wL$	$0.414wL$

— for special case when $L_C = 0.172L$

For maximum positive moment equal to maximum negative moment, $L_C = 0.172L$. For this special case, the maximum deflection occurs in span between R_1 and R_2 and is

$$\Delta_{\max} = \left(\frac{0.370}{48}\right)\left(\frac{wL^4}{EI}\right)$$

(Note: The "x" subscript for M and V denotes moment and shear, respectively, for the span between R_1 and R_2. The "y" subscript for M and V is for the span between R_2 and R_3.)

7. Three Equal Spans: Uniform Load

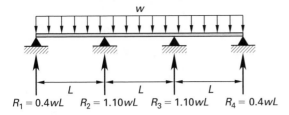

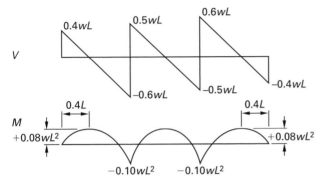

$$\Delta_{\text{max}} = \frac{wL^4}{145EI} \text{ at } 0.446L \text{ from } R_1 \text{ or } R_4$$

8. Continuous Beam: Three Equal Spans (One End Span Unloaded)

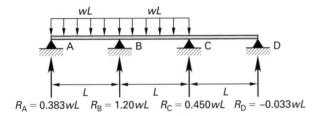

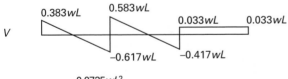

$$\Delta_{\text{max}} \ (0.430L \text{ from A}) = \frac{0.0059wL^4}{EI}$$

9. Continuous Beam: Three Equal Spans (End Spans Loaded)

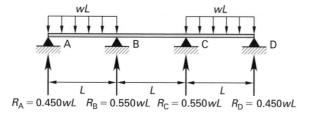

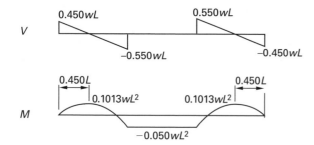

$$\Delta_{\max} \ (0.479L \text{ from A or D}) = \frac{0.0099wL^4}{EI}$$

References

American Association of State Highway and Transportation Officials, Inc. (AASHTO), *Standard Specifications for Highway Bridges*, 16th ed., Washington, D.C., 1996, with 1998 Interims.

American Forest & Paper Association, *ANSI/AF&PA NDS-1997, National Design Specification for Wood Construction*, 1997.

American Forest & Paper Association, *ANSI/AF&PA 1997 Edition Supplement, National Design Specification*, 1997.

American Institute of Timber Construction, *Timber Construction Manual*, 4th ed., New York: John Wiley & Sons, 1994.

American Plywood Association (APA), *Design/Construction Guide: Diaphragms and Shear Walls*, Tacoma, WA, 1997.

American Plywood Association (APA), *Plywood Design Specification*, Tacoma, WA, 1998.

American Plywood Association (APA), *PDS Supplement 2: Design and Fabrication of Plywood-Lumber Beams*, Tacoma, WA, 1992.

American Society of Civil Engineers, *Minimum Design Loads for Buildings and Other Structures*, ASCE 7-95, ASCE STANDARD, New York, 1996.

Breyer, Donald E., K.J. Fridley, and K.E. Cobeen, *Design of Wood Structures*, 4th ed., New York: McGraw-Hill, 1999.

Faherty, Keith F., and Thomas G. Williamson (editors), *Wood Engineering and Construction Handbook*, 3rd ed., New York: McGraw-Hill, 1999.

Forest Products Laboratory, Forest Service, U.S. Department of Agriculture, *Wood Handbook: Wood as an Engineering Material*, Forest Products Society, 2801 Marshall Court, Madison, WI, 1999.

Guinther, J.A., J.B. Kim, and S. Cabler, *Stiffness Evaluation of Metal Plates Connected on Wood Members*, Proceedings of 5th World Conference on Timber Engineering, Montreux, Switzerland, 254–260, 1998.

Gurfinkel, German, *Wood Engineering*, 2nd ed., Dubuque, IA: Kendall/Hunt, 1981.

Kim, Jai B., and Chris Brill-Edwards, "Comparison of Full-Scale Truss Test Results with Those by Analytical Methods," The First RILEM Symposium on Timber Engineering Proceedings, Stockholm, Sweden, 1999, pp. 119–128.

Kim, Jai B., and Kevin Barron, "Non-Linear Modeling of the Heel Joint of Metal Plate Connected Roof Trusses," The World Conference on Timber Engineering 2000, Whistler Resort, British Columbia, Canada, 2000, pp. 7.3.4-1–7.3.4-10.

Kim, Robert H., and Jai B. Kim, *Effectively Spliced Large Timbers and Full-Scale Tests on Spliced Timbers*, Proceedings of 5th World Conference on Timber Engineering, Montreux, Switzerland, 798–799, 1998.

Kim, Robert H., and Jai B. Kim, "Oak A-Frame Timber Bridges Meeting the Modern Deflection Requirement," Transportation Research Record No. 1319, Washington D.C., 1991.

Smith, R.L., and J.B. Kim, *Life Cycle Cost Considerations for Timber Bridges*, Proceedings of Structures Congress, ASCE, Portland, Oregon, 233–237, 1997.

Stalnaker, Judith J., and Earnest C. Harris, *Structural Design in Wood*, 2nd ed., New York: Chapman & Hall, 1997.

Index

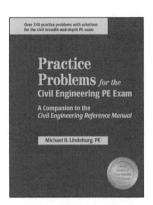
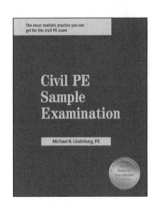
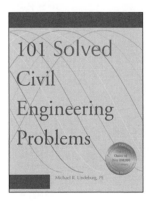

Turn to PPI for All of Your Exam Preparation Needs

PPI is your one stop for review manuals, practice problems, sample exams, quick references, and much more!

Visit www.ppi2pass.com to see our complete selection of review materials for the FE and PE exams.

FE Exam Review

FE Review

FE Review Manual

EIT Review

Engineer-In-Training Reference Manual

EIT Solutions

Solutions Manual for the Engineer-In-Training Reference Manual

Sample Exams

FE/EIT Sample Examinations

Civil PE Exam Review

Reference Manual

Civil Engineering Reference Manual for the PE Exam

Practice Problems

Practice Problems for the Civil Engineering PE Exam

Sample Exam

Civil PE Sample Examination

Quick Reference

Quick Reference for the Civil Engineering PE Exam

Mechanical PE Exam Review

Reference Manual

Mechanical Engineering Reference Manual for the PE Exam

Practice Problems

Practice Problems for the Mechanical Engineering PE Exam

Sample Exam

Mechanical PE Sample Examination

Quick Reference

Quick Reference for the Mechanical Engineering PE Exam

Electrical PE Exam Review

Reference Manual

Electrical Engineering Reference Manual for the PE Exam

Practice Problems

Practice Problems for the Electrical and Computer Engineering PE Exam

Sample Exam

Electrical and Computer PE Sample Examination

Quick Reference

Quick Reference for the Electrical and Computer Engineering PE Exam

Environmental PE Exam Review

Reference Manual

Environmental Engineering Reference Manual

Practice Problems

Practice Problems for the Environmental Engineering PE Exam

Practice Exams

Environmental Engineering Practice PE Exams

Chemical PE Exam Review

Reference Manual

Chemical Engineering Reference Manual for the PE Exam

Practice Problems

Practice Problems for the Chemical Engineering PE Exam

Practice Exams

Chemical Engineering Practice PE Exams

Quick Reference

Quick Reference for the Chemical Engineering PE Exam

Structural PE Exam Review

Reference Manual

Structural Engineering Reference Manual for the PE Exam

Solved Problems

246 Solved Structural Engineering Problems

Order today!
Visit www.ppi2pass.com
or call 800-426-1178.

Professional Publications, Inc.

www.ppi2pass.com

Promotion Code: **EBIS**